Sabine König

Probing the ULIRG-to-QSO evolution

Sabine König

Probing the ULIRG-to-QSO evolution

Properties of Quasistellar Objects and their progenitors in the millimeter and radio wavelength regimes

Südwestdeutscher Verlag für Hochschulschriften

Impressum/Imprint (nur für Deutschland/ only for Germany)
Bibliografische Information der Deutschen Nationalbibliothek: Die Deutsche Nationalbibliothek verzeichnet diese Publikation in der Deutschen Nationalbibliografie; detaillierte bibliografische Daten sind im Internet über http://dnb.d-nb.de abrufbar.

Alle in diesem Buch genannten Marken und Produktnamen unterliegen warenzeichen-, marken- oder patentrechtlichem Schutz bzw. sind Warenzeichen oder eingetragene Warenzeichen der jeweiligen Inhaber. Die Wiedergabe von Marken, Produktnamen, Gebrauchsnamen, Handelsnamen, Warenbezeichnungen u.s.w. in diesem Werk berechtigt auch ohne besondere Kennzeichnung nicht zu der Annahme, dass solche Namen im Sinne der Warenzeichen- und Markenschutzgesetzgebung als frei zu betrachten wären und daher von jedermann benutzt werden dürften.

Verlag: Südwestdeutscher Verlag für Hochschulschriften Aktiengesellschaft & Co. KG
Dudweiler Landstr. 99, 66123 Saarbrücken, Deutschland
Telefon +49 681 37 20 271-1, Telefax +49 681 37 20 271-0
Email: info@svh-verlag.de
Zugl.: Köln, Universität, Inauguraldissertation, 2009

Herstellung in Deutschland:
Schaltungsdienst Lange o.H.G., Berlin
Books on Demand GmbH, Norderstedt
Reha GmbH, Saarbrücken
Amazon Distribution GmbH, Leipzig
ISBN: 978-3-8381-1784-3

Imprint (only for USA, GB)
Bibliographic information published by the Deutsche Nationalbibliothek: The Deutsche Nationalbibliothek lists this publication in the Deutsche Nationalbibliografie; detailed bibliographic data are available in the Internet at http://dnb.d-nb.de.

Any brand names and product names mentioned in this book are subject to trademark, brand or patent protection and are trademarks or registered trademarks of their respective holders. The use of brand names, product names, common names, trade names, product descriptions etc. even without a particular marking in this works is in no way to be construed to mean that such names may be regarded as unrestricted in respect of trademark and brand protection legislation and could thus be used by anyone.

Publisher: Südwestdeutscher Verlag für Hochschulschriften Aktiengesellschaft & Co. KG
Dudweiler Landstr. 99, 66123 Saarbrücken, Germany
Phone +49 681 37 20 271-1, Fax +49 681 37 20 271-0
Email: info@svh-verlag.de

Printed in the U.S.A.
Printed in the U.K. by (see last page)
ISBN: 978-3-8381-1784-3

Copyright © 2010 by the author and Südwestdeutscher Verlag für Hochschulschriften Aktiengesellschaft & Co. KG and licensors
All rights reserved. Saarbrücken 2010

Meiner Familie

Contents

0.	**Overview**	**1**
1.	**Activity in galactic nuclei**	**7**
	1.1. Introduction	7
	1.2. Classification of AGN	8
	1.2.1. Quasars and QSOs	9
	1.2.2. Blazars	10
	1.2.3. Seyfert galaxies	11
	1.2.4. LINERs	13
	1.2.5. ULIRGs	13
	1.3. Emission lines as diagnostic tracers of physical properties	15
	1.3.1. Atomic gas	16
	1.3.2. Molecular gas	17
	1.3.3. Maser	20
2.	**Observations in the radio and mm wavelength regime**	**23**
	2.1. Single-dish telescopes	24
	2.1.1. The Effelsberg 100-m telescope	26
	2.2. Interferometry and interferometers	27
	2.2.1. The Very Large Array (VLA)	29
	2.2.2. The IRAM Plateau de Bure Interferometer	30
	2.3. Observation methods	31
	2.3.1. Position-switching	31
	2.3.2. Load-switching	32

3. The Arp 220 merger on kpc scales — 35

- 3.1. Introduction 35
- 3.2. Observations and data reduction 37
- 3.3. Unveiling the central kiloparsecs 37
  - 3.3.1. Arp 220-West 39
  - 3.3.2. Arp 220-East 39
- 3.4. The different kinematic scales in Arp 220 41
- 3.5. The large scale picture 43
- 3.6. Identikit simulations 45
  - 3.6.1. The Identikit model 45
  - 3.6.2. Simulation results 46
  - 3.6.3. Discussion 52
- 3.7. Conclusions 55

4. HI in nearby low-luminosity QSO host galaxies — 57

- 4.1. Introduction 57
- 4.2. The sample 58
- 4.3. Observations and data reduction 60
- 4.4. Analysis 64
- 4.5. Results and discussion 67
  - 4.5.1. Source confusion 72
  - 4.5.2. Morphology 73
  - 4.5.3. IRAS color-color diagram 78
  - 4.5.4. Statistical considerations 79
- 4.6. Follow-up VLA observations 81
  - 4.6.1. Observations and data reduction 81
  - 4.6.2. Results 81
- 4.7. Conclusions 82

5. A search for H_2O maser emission in nearby low-luminosity QSO host galaxies — 85

- 5.1. Introduction 85
- 5.2. The sample 86

	5.3.	Observations and data reduction	87
	5.4.	Results/Discussion	89
		5.4.1. Sensitivity	92
		5.4.2. Host galaxy properties	95
		5.4.3. Morphology	95
		5.4.4. Black hole mass	98
	5.5.	Conclusions/Summary	99

A. Appendix 101

A.1. The H_2O maser spectra and DSS2 IR images 101

A.2. The H_2O maser spectra and the corresponding cross-correlation plots 107

Bibliography 134

List of Figures 136

List of Tables 137

0. Overview

This thesis tries to shed some light on QSOs and the properties of their hosting galaxies in the radio and mm wavelength domains. Atomic and molecular gas makes up a significant part of the content of the material in a galaxy. Previous investigations of radiation emitted by the gaseous galaxy content of extragalactic objects show very different physical properties for very different types of objects. But also objects with very similar properties/appearances were found. Examples in the context of this thesis are, e.g., ULIRGs and QSOs, which can be studied in to the so-called ULIRG-to-QSO evolutionary scheme: The hypothesis that ULIRGs are the dust enshrouded progenitors of QSOs (Sanders et al., 1988). In this context objects from both, the intermediate and the late phase of this evolutionary path, are presented here in the form of Arp 220, as one prototypical ULIRG, and the nearby QSO sample. Both, the atomic and the molecular gas properties of their hosting galaxies are studied in this thesis. The study of Arp 220 is focused on the molecular properties of the merging galaxy. In addition, to that the merger in Arp 220 is modeled by a simulation tool using a collisionless N-body code. The results on the nearby QSO sample present an overview over the neutral atomic and the molecular gas properties of the host galaxies. Neutral atomic hydrogen and water vapor megamaser emission are used as tracers for the properties of the galactic contents.

In Chapters 1 and 2 an introduction to the most important fundamentals of this thesis is given. Chapter 1 summarizes the basic comprehension of the different classes of AGN and gives a brief introduction to radio and mm astronomical tracers, used for the observation of the afore mentioned objects.

Overview

In Chapter 2 the observational techniques used for the data obtainment, for the studies presented here, are introduced. A short overview of the three main chapters is given in the following.

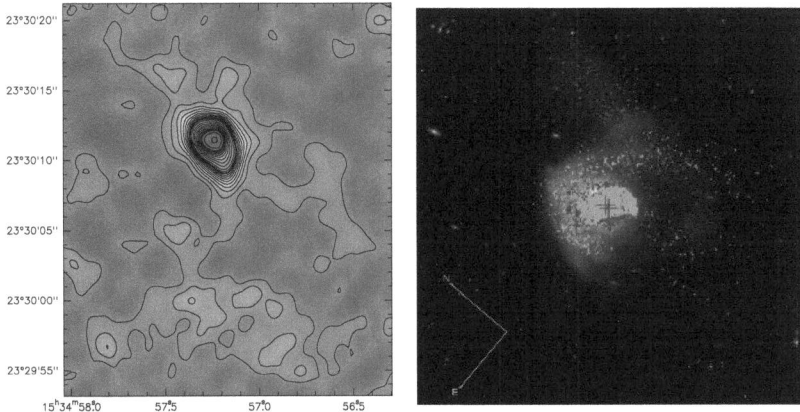

Integrated intensity map of the CO(1–0) emission in Arp 220 *left* and results for the best model fit, with an HST composite image of Arp 220 in the background (*right*).

Chapter 3: The Arp 220 merger on kpc scales

Here the eastern nucleus of Arp 220 is studied and the whole galaxy was searched for the more extended molecular gas in different CO line transitions of the famous ULIRG. Using the PdBI, CO (2-1) and (1-0) data at 1 and 3 mm were obtained in 1994, 1996, 1997 and 2006 at different beam sizes and spatial resolutions. The observations show that the SE (south-east) peak is most likely the 'real', but deeply embedded/highly dust obscured, nucleus of Arp 220-East. Moreover, Arp 220-East shows a rotational motion that corresponds to the overall rotation of the underlying molecular disk. On the position of CO-SW (south-west) a peak in the line width distribution and the rotational motion in the velocity field indicates the probable presence of an additional, very faint nucleus. For the first time it is tried to detect and study the molecular gas outside the central $3''$. There are indications, in the CO (1-0) low resolution maps, for emission about $10''$ towards the south, as well as to the north and to the west of the nuclei. Furthermore the merger in Arp 220 was modeled with the help of the CO data and an HST image of the prototypical ULIRG. The simulations of the merger in Arp 220 were performed with the Identikit modeling tool. The model parameters that describe the galaxy merger best give a mass ratio of 1:2 and result in a merger of $\sim 6 \times 10^8$ yrs.

Overview

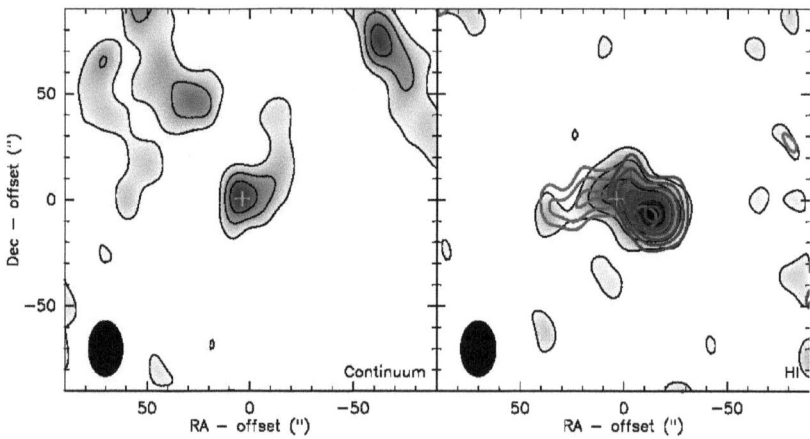

Atomic gas content of the nearby QSO host galaxy HE 1248–1356. The continuum map is shown on the (*left*), the HI line emission map on the (*right*).

Chapter 4: HI in nearby QSO host galaxies

This chapter addresses the atomic gas properties of a sample of nearby low-luminosity QSO host galaxies. 27 previously CO detected sources were searched for HI emission. The relationship between the HI and infrared properties of the host galaxies is investigated. In addition, the atomic and molecular gas contents were compared and the optical and FIR (far-infrared) properties were searched for connections. The single-dish observations were made with the Effelsberg 100-m telescope. All sample objects were drawn from the Hamburg/ESO survey of optically bright QSOs with declinations $\delta > -30°$ and redshifts up to $z = 0.06$. 12 host galaxies from the sample have been detected in the HI 21 cm emission line. The median HI gas mass in the whole sample is of the order of $11.4 \times 10^9 \, M_\odot$, which is a factor of two higher than the HI content of the Milky Way, and it corresponds to the upper limit of HI masses found for ULIRGs. The data points of the non-detected sources sample an upper envelope of the HI mass distribution.

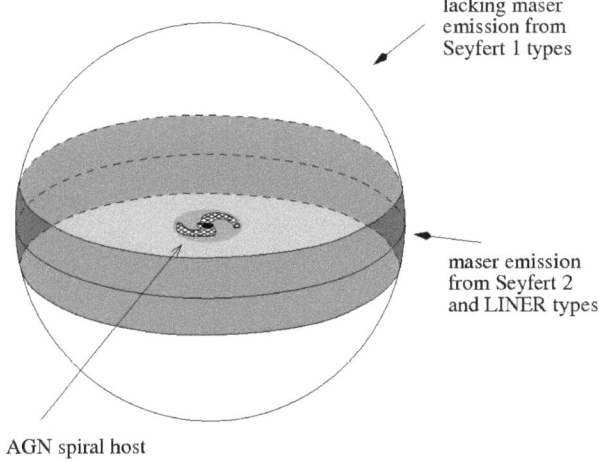

A simple depiction of the unified scheme in context of water megamaser emission.

Chapter 5: Water megamasers in nearby QSOs?

In this part of the thesis, the study of the sample of nearby QSO host galaxies is continued. Using the Effelsberg 100-m telescope again, the 17 most IR (infrared) luminous sources were searched for emission in the 22 GHz water vapor maser transition. None of the observed sources was detected in water maser emission. The discussion on the reasons for H_2O megamasers being rarely found in Seyfert 1 galaxies are reviewed. Furthermore the differences between Seyfert galaxies and QSOs are discussed, as well as possible reasons for the absence of water megamaser emission in the sample QSOs. Eight of the observed sources show a bulge dominated morphology and are probably of elliptical type (E/S0), whereas 6 have a spiral geometry. Three of the objects seem to be in a late phase of merging/interaction. $3\,\sigma$ upper limits, for the flux density, of 27 to 60 mJy at spectral resolutions of ~ 0.43 km s^{-1} were found.

1. Activity in galactic nuclei

1.1. Introduction

Approximately 60 years ago, a very 'special' class of astronomical object was discovered: Optically, their appearance is point-like and it was proposed that these objects are a new unusual phase in the evolution of stars. But the spectra didn't look star-like at all. The distances for these sources, determined by optical spectroscopy, shed some light on their unusual nature. They turned out to be located outside the Milky Way at high redshifts (3C 273, $z=0.158$; 3C 48, $z=0.367$ Schmidt, 1963; Greenstein & Schmidt, 1964). Due to their appearance and extragalactic distances, this class of objects was termed quasi-stellar radio sources (quasars) or quasi-stellar objects (QSOs), depending on their properties in the radio waveband. But that was not the only peculiar behavior observed for these objects: Resulting from the findings about their extragalactic origin it was discovered that almost all the energy emitted by quasars originated from the inner central regions of the galaxy, the so-called active galactic nucleus (AGN). Therefore the point-like appearance was explained by a galaxy which has its nucleus outshining the surrounding environment of the fainter host. The spectral energy distribution (SED) of these objects extends over a wide range of the electromagnetic spectrum with roughly equal energy emitted in each waveband. The nature of the mechanism responsible for the huge energy output, that cannot be explained by pure stellar radiation, is not clear to date. The most likely explanation to account for the nuclear activity is the process of gravitational accretion of matter by a supermassive black hole (SMBH): Magorrian et al. (1998) found a relation between the black hole mass and the mass of the bulge of a galaxy. In combination with the correlation between the black hole mass and the stellar velocity dispersion (e.g., Gebhardt et al., 2000; Shields et al., 2003), this implies that most galaxies with a bulge component harbor a supermassive black hole in their center (see e.g., Richstone et al., 1999; Kormendy & Gebhardt, 2001). The only other real contender, for the supermassive black hole as the responsible component for activity in the galactic nucleus, is the nuclear starburst scenario (see e.g., Terlevich et al., 1992). All other alternative models have been rejected with time due to the incapability of the models to account for several AGN characteristics.

One central question, however, is still unsolved: How is the AGN fueled? One hypothesis is that AGN activity may be triggered by interactions with other galactic systems. Mergers are a very efficient way for matter to loose angular momentum and to be transported to the central regions of the galaxy, within the vicinity of the black hole. But not each galaxy with an AGN has close-by companions or shows any sign of interactions/mergers in its past. Hence the study of the AGN's environment, starting with the hosting galaxy, is essential in the search for answers to the unsolved questions. Studying the host galaxy itself is one approach to gather knowledge about the true nature of AGN. Due to their small cosmological distance, nearby QSOs are ideally suited to study the properties of their host galaxies. Important diagnostic tools to explore the physical mechanisms and properties in these objects, on very different scales, are the atomic and molecular gas species.

1.2. Classification of AGN

Most AGN systems are very similar in their physical properties. Some measures for classifications are the luminosities, the presence of jets, outflows and other morphological peculiarities or anisotropic effects (Binney & Merrifield, 1998). The first active galaxies discovered looked like stars but showed a non-stellar spectrum. They were named quasars (quasi-stellar radio source) due to their strong radio emission. Later discovered quasars with weaker radio emission were termed QSOs (quasi-stellar objects). Blazars are objects with luminosities comparable to those of QSOs, but are highly variable. A similar class of galaxies with lower luminosities (factor of ~ 100 compared to QSOs/quasars) are the Seyfert galaxies. Due to their fainter nuclei the host galaxies of Seyfert nuclei are visible, in contrast to quasars. Continuing the series to objects of even smaller luminosities, LINERs (*Low Ionization Nuclear Emission Regions*) complete the overview. Unification models (e.g. Antonucci, 1993; Urry & Padovani, 1995) are means to study the evolution of galaxies, and QSOs, in particular.

1.2.1. Quasars and QSOs

Quasars (quasi-stellar radio sources), the subclass of AGN with the highest luminosities, were discovered as unresolved, starlike objects with a clear non-stellar spectrum. Their starlike appearance originates in their optically exceptionally bright active galactic nuclei (AGN), which dominate the brightness distribution of the quasar host galaxies. A widely accepted explanation for the quasars strong emission at all wavelengths is the accretion of matter onto a supermas-

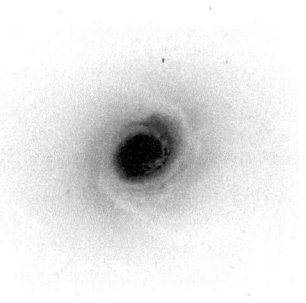

Figure 1.1.: The type 1 QSO NGC 4593 (taken from the NED). Reproduced by permission of the AAS.

sive black hole. Quasars are very similar to the group of objects summarized as type 1 Seyfert galaxies. In fact, the only difference in the classification between these two object classes is made with the help of the absolute magnitude M_B: Objects fainter than absolute magnitude values of -23 mag ($M_B > -23$ mag) are termed Seyfert galaxies. Objects brighter than this limit ($M_B \leq -23$ mag) are quasars. Quasars occupying the lower end of the luminosity function found for quasars are termed low-luminosity quasars. Furthermore, an additional distinction within the group of quasars was introduced: Radio-loud objects are classified as quasars whereas radio-quiet quasars are classified as QSOs (*q*uasi-*s*tellar *o*bjects)[1]. Quasars are usually found at higher redshifts (their number density peaks at redshifts $2 \leq z \leq 3$, e.g., Schmidt, 1988). There are indications that quasars result from interacting/merging galaxy systems (e.g. Sanders et al., 1988; Bahcall et al., 1997).

1.2.2. Blazars

Under the term blazars, the groups of BL Lac objects and OVVs (Optically Violently Variables), very luminous objects, are united. BL Lac objects, named after their prototype BL Lacertae, are galaxies exhibiting only weak emission lines. They show luminosities comparable to QSOs, but their spectra are featureless (Ghosh et al., 2000). These types of objects have a very compact morphology, i.e. mostly elliptical morphology. They are most well-known for their highly variable emission at almost all wavelengths. OVVs however are objects with strong emission lines, that show equally strong flux variabilities in the optical wavelength regime (Peterson, 1997, and references therein).

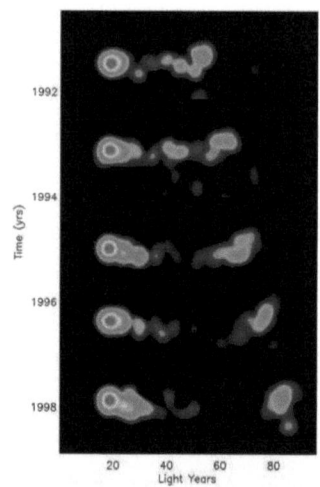

Figure 1.2.: Superluminal motion in the blazar 3C 279 is shown in a mosaic of five radio images made over seven years (Image courtesy of NRAO/AUI/NSF, http://images.nrao.edu/387).

Within the unified scheme blazars are believed to be AGNs with a powerful relativistic jet close to the line-of-sight to the observer. Therefore blazars are often associated with superluminal motions.

---

[1] A standard division between radio-loud and radio-quiet is given by the radio power $P_{5\,GHz}$, $\log_{10}[P_{5\,GHz}(W/Hz/sr)] = 24$ (Goldschmidt et al., 1999)

1.2.3. Seyfert galaxies

Seyfert (Sy) galaxies belong to the group of low-luminosity AGN. They show luminosities a factor of ~ 100 lower than for QSOs/quasars. Seyfert galaxies, or short Seyferts, are characterized by very fast motions of the gas resulting in the broadening of emission lines to line widths of ~ 500 up to $10\,000$ km s^{-1}. Their highly excited, strong emission lines, in the infrared wavelength regime, originate from their nuclei. Most Seyfert nuclei reside in host galaxies

Figure 1.3.: The type 1 Seyfert galaxy NGC 7742 (from *http://imgsrc.hubblesite. org/hu/db/images/hs-1998-28-a-full_jpg. jpg, credit: The Hubble Heritage Team (AURA/STScI/NASA)*).

with a spiral morphology, but there are also a few exceptions for the host galaxies being of elliptical structures. Due to the large spread in line widths a further subclassification was introduced: Type 2 Seyfert galaxies only show narrow lines (500 to $1\,000$ km s^{-1}) whereas type 1 Seyfert galaxies show very broad lines (Δv up to $10\,000$ km s^{-1}) but also narrow emission lines. The narrow line component in Seyfert 1 galaxies is made up of, e.g., forbidden line transitions of oxygen (e.g., [OII], [OIII]). The classification into type 1 and 2 Seyfert galaxies has no 'hard limit'. However, observations of objects characterized as Seyfert galaxies also show intermediate line widths in their spectra. Hence another subclassification was introduced: Dependent on the line widths found in the object's spectrum the galaxy is classified as e.g., Seyfert type 1.5, 1.8, 1.9 up to Seyfert type 2. One major difference between the spectra of Seyfert 1s and Seyfert 2s was found, as mentioned before, in the line widths of their emission lines. The comparison of Seyfert 1 galaxy spectra to QSOs yields that spectra of Seyfert 1 galaxies have, on average, a similar spectrum as quasars. However, due to the large line widths in Seyfert 1s, on first sight there is no resemblance between the type 1 and 2 Seyfert galaxies. Later it was found that in polarized light the spectra of Seyfert 2s and Seyfert 1s are indeed quite similar to each other (e.g. Krolik

1. Activity in galactic nuclei

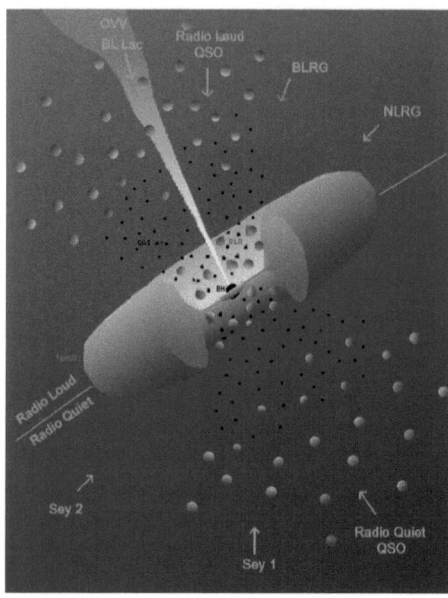

Figure 1.4: This cartoon shows the main features of a 'unified model' for AGN as reviewed by Urry & Padovani (1995). Depending on the viewing angle of the observer the galaxy shows an appearance of a Seyfert 1 (Sey 1) or Seyfert 2 (Sey 2) galaxy, or a radio-quiet QSO. Radio-loud galaxies on the other hand can have the appearance of OVVs or BL LACs, radio-loud QSOs or as broad-line (BLRG) and narrow-line radio galaxies (NLRG).

& Kallman, 1987; Wilkes et al., 1995). This finding led to the introduction of the so-called unified scheme of the different Seyfert type galaxies. In Fig. 1.4, a cartoon of this scenario is shown. It says that Seyfert type 1 and 2 galaxies only differ in terms of the viewing angle of the observer. In other words: Seyfert 1 nuclei are hidden within Seyfert 2 galaxies behind an obscuring dusty molecular thick disk or torus. The studied objects will appear as Seyfert 1s when the observer has a more pole-on view on the nucleus, whereas they will appear as Seyfert 2 galaxies with a more edge-on view. In terms of the characterization of the Seyfert type by line width this means the following: The Seyfert 1 nuclei, together with the broad line (BLR) and the narrow line regions (NLR) are directly visible and not significantly obscured by material. The emission from the broad line region (BLR) in Seyfert 2 nuclei is covered by the dusty torus. Hence, from the nuclei of type 2 Seyfert galaxies only the light reflected by dust grains reaches the observer. This reflected light, which is generally polarized, leads to the similarity of Seyfert 1 and 2 polarized emission spectra.

1.2.4. LINERs

A group of galaxies even less luminous than Seyfert 2 galaxies are the so-called *Low Ionization Nuclear Emission Regions* (LINERs). They were first discovered by Heckman (1980), using oxygen lines. Spectroscopically, LINER galaxies are very similar to type 2 Seyfert galaxies but can be distinguished from them on the basis of emission line ratios (e.g., [OIII]$\lambda 5007$/Hβ, [NII]$\lambda 6583$/Hα). Seyfert 2s have high values in both ratios, whereas for LINERs the values of [OIII]$\lambda 5007$/Hβ are relatively low in comparison to their [NII]$\lambda 6583$/Hα ratio. Spectra of typical LINER galaxies show emission from weakly ionized or neutral atoms (e.g., O, OI, NI). Approximately $\sim 30\%$ of all nearby galaxies are classified as LINER galaxies (distances up to 20-40 Mpc Heckman, 1980). A still ongoing debate is whether the ions in LINERs are excited by shock waves, propagating through the gaseous material (Heckman, 1980), or by photoionization (Ho et al., 1993). Together with the Seyfert 2s, LINERs make up the majority of the population of galaxies known to exhibit emission in the 22 GHz water megamaser line transition (Braatz et al., 1997; Kondratko et al., 2006b).

Figure 1.5.: The LINER galaxy M 51 (http://hubblesite.org/gallery/album/entire/pr2005012a/1280_wallpaper/, credit: ESA, NASA, S. Beckwith (STScI), and The Hubble Heritage Team (STScI/AURA)).

1.2.5. ULIRGs

Already in 1972, Rieke & Low mentioned the existence of galaxies with an IR excess. These sources, called *U*ltra*L*uminous *I*nfra*R*ed *G*alaxies (ULIRGs) are objects of a galaxy class characterized by very high luminosities in the infrared wavelength regime discovered by the IRAS satellite (Infrared astronomy satellite). Their IR luminosity ranges between $10^{12} L_\odot \leq L_{\text{bol}} \sim L_{\text{IR}}$ [$8 - 1\,000\,\mu$m] $\leq 10^{13} L_\odot$. In the optical, UV and even in the mid-infrared and X-ray wavelength bands ULIRGs are highly obscured which makes it very difficult to clearly probe the physical properties by observations. They are so bright in the infrared because this obscuring dusty material re-radiates the energy from the power source of the ULIRGs. They are thought to be powered by either a strong nuclear starburst (Joseph, 1999), a heavily obscured active galactic nucleus (Sanders, 1999) or a mixture of the two. Whatever the power source may be, a large molecular gas concentration in the central kiloparsecs is needed in order to be sustained (see e.g., Sanders et al., 1988).

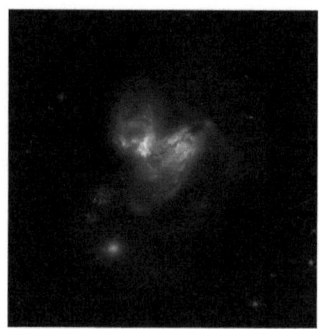

Figure 1.6.: The ULIRG Arp 299 (http://hubblesite.org/gallery/album/entire /pr2008016bs/xlarge_web/, credit: NASA, ESA, the Hubble Heritage (STScI/AURA)-ESA/Hubble Collaboration, and A. Evans (University of Virginia, Charlottesville/NRAO/Stony Brook University)).

Most ULIRGs are found to be in a state of merger or interaction with other galaxies (e.g. Farrah et al., 2001; Veilleux et al., 2002). Previous studies of the molecular gas content of ULIRGs show that tracers indicating high-density gas originates in a compact nuclear region (see e.g. Sanders et al., 1991; Gao & Solomon, 2004). Hou et al. (2009) state that $\sim 25\%$ of all ULIRGs show indications for the presence of AGN (~ 15–20% according to Nardini et al. (2008)). Even more so (about 50%), if the infrared luminosity of the observed ULIRG surpasses a value of $10^{12.3}$ L$_\odot$. Sanders et al. (1988) found that ULIRGs and QSOs have similar properties, such as similar space densities (Soifer et al., 1986) and luminosities. Hence they proposed a model in which ULIRGs play a dominant role in the formation of all QSOs. The scenario indicates that ULIRGs are an early phase, dust-enshrouded, object on the evolutionary road to the formation of QSOs: Two spiral galaxies, very rich in molecular gas, merge or interact. During the merger/strong interaction the decrease of orbital angular momentum results in a decrease of the orbital radius. This decrease in angular momentum then further results in the funneling of molecular gas to the central regions of the merging galaxies. This enhancement of gaseous material will result in an intense burst of star formation. At the same time this gas, and stellar remnants introduced by the strong interactions provide the fuel for an active galactic nucleus. At this point in the evolution the far-infrared luminosity is dominated by the starburst contribution. The time period to the maximum contribution of the starburst component is assumed to be $\sim 10^8$ yrs. The increase in the contribution of the AGN generally takes place when the merging nuclei reach apparent separations less than ~ 1 kpc. In the ULIRG phase the starburst is still switched on but is then overpowered by the AGN contribution (Veilleux et al., 2009). The combination of the radiation pressure and super novae explosions results in the blowing out of the dust. From now on the objects will take on the appearance of classical QSOs.

Kawakatu et al. (2006) and Cao et al. (2008) searched for the link between ULIRGs and QSOs. They propose that type 1 ULIRGs, ULIRGs containing Seyfert 1 nuclei, are the missing link in the evolutionary scenario. Veilleux et al. (2009), however, found in their bigger sample that there are fully merged ULIRGs which have not yet produced systems dominated by the AGN contribution or which have not blown away the obscuring dust envelope. They also find ULIRGs in a pre-merger state that are already dominated by AGN contribution. They conclude that the scenario presented by Sanders et al. (1988) needs to be revised and replaced by a 'softer' version of the evolutionary scenario: Veilleux et al. (2009) propose a scenario where several evolutionary paths are possible during the period of AGN contribution, i.e. allowing for non-monotonic episodes in the evolution in quantities like AGN contribution to the luminosity, Eddington ratio, etc.

1.3. Emission lines as diagnostic tracers of physical properties

In this thesis, atoms and molecules are used as tracers of conditions and physical processes in external galaxies. Depending on kinetic temperature and density of the gaseous material different atoms and molecules are employed to study the physical properties of the galaxies. The observed line spectra can be used to determine densities, temperatures and masses. Furthermore statements on the kinematics and dynamics of the host galaxies can be made.

1.3.1. Atomic gas

The neutral atomic gas consists of cold diffuse neutral hydrogen (HI) clouds and warm inter cloud gas. The far ultra-violet (FUV) field around stars is the energy source for the radiation emitted by the atomic gas, it provides the energy needed to excite the atomic line and continuum emission. Within the clouds the gas has a filamentary structure. Observations of the neutral atomic gas are done in emission as well as with absorbtion lines. The cold neutral medium corresponds to a cold, dense cloud phase. It fills about 2.5% of the volume of the interstellar medium (ISM). This material is studied by observations of the 21 cm HI emission line and interstellar absorption lines. The warm neutral medium, on the other hand, makes up 15% of the ISM volume. For studies of the warm neutral medium, diffuse X-ray background emission and absorbtion lines in the UV (ultra-violet) regime, are used. The filling factors stated here are based on Galactic measurements.

1.3.1.1. Neutral atomic hydrogen

Neutral atomic hydrogen emission at a wavelength of 21 cm (1.4 GHz) is radiated via magnetic dipole radiation, originating from the hyperfine transition of the ground state of atomic hydrogen. The transition between the energy levels involved takes place between the hyperfine doublet ($F = 1 \rightarrow F = 0$) of the ground state of the atomic hydrogen.

In 1951, Ewen & Purcell were the first to observe the 21 cm line transition of neutral atomic hydrogen in our galaxy. Nowadays this line transition is used for a variety of astrophysical purposes: e.g., the determination of the total neutral gas mass of galaxies, the rotation curve of galaxies and hence also their atomic gas structure and the dynamics of the individual galaxies. In this context HI can be used as a tracer of the gravitational potential, produced by the total galaxy content, and hence of the dynamical state of the galaxy. With the help of the rotation curves and velocity fields deduced from the HI data the potentials of such galaxies are determined. It was discovered that for 'normal' galaxies, without any peculiarities, the dimensions for the HI gas are much larger than, e.g., for the molecular gas or the stellar components of the galaxies. Because of its greater extension in the host galaxies, with respect to molecular gas (e.g. CO), HI is especially suitable for the search for early stages of interaction with the surrounding medium (e.g., in the M 81 galaxy group Kuo et al., 2008). The shapes of asymmetric HI spectra can therefore be used as indicators for asymmetries on larger scales. These can then be studied with respect to the subject of massive mergers in nearby QSO host galaxies. The extended emission is furthermore used to trace the system kinematics, dynamics and provides evidence for the presence of dark matter in the outer galaxy regions (e.g., Hewitt et al., 1983; Martin, 1998; Bajaja et al., 1994).

Only little is known so far on the atomic gas content in QSO host galaxies. Lim & Ho (1999) and Lim et al. (2001) present one of the first studies of HI emission in nearby QSO host galaxies. They find that while most of their observed sources show signs of a past or ongoing tidal interaction, they do not yet physically merge with another galaxy. An extensive survey of 101 northern nearby galaxies hosting type 1 AGN was conducted by Ho et al. (2008a,b). Their sample shows a detection rate of $\sim 65\%$. They find a strong relation between the mass of the central black hole and the dynamical mass of the host galaxy. Furthermore, Ho et al. (2008b) present evidence for a substantial ongoing black hole growth in the most actively accreting AGNs in their sample. Haan et al. (2008) present a high resolution study of 16 nearby NUGA (*NU*clei of *GA*laxies) sources of different AGN types.

The observations of HI presented in this thesis are used to study the atomic gas content in nearby QSO host galaxies.

1.3.2. Molecular gas

Molecular gas shows complex structures like clumps and clouds whose densities range from $n \sim 10^5$ up to 10^9 cm^{-3}. With a mass fraction, of the whole mass content of a galaxy, of 10% the molecular gas occupies only 1% of

1. Activity in galactic nuclei

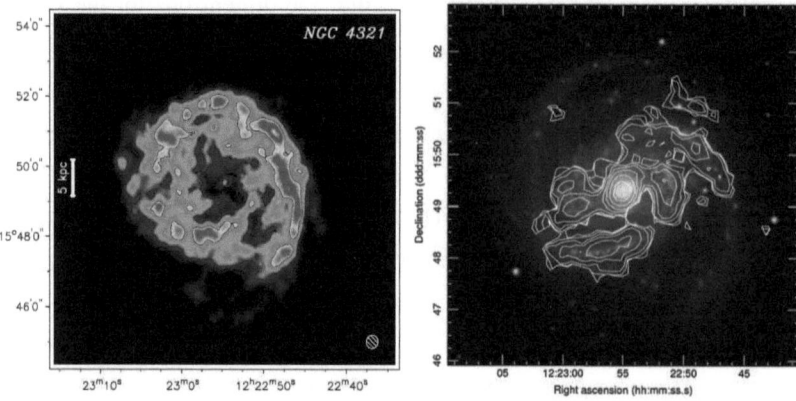

Figure 1.7.: NGC 4321 in HI (*left* Haan et al., 2008, reproduced by permission of the AAS.) and in ^{12}CO(3−2) line emission (*right*, on top of a DSS image, from the JCMT Legacy Survey, http://outreach.jach.hawaii.edu/pressroom/2008_legacysurvey1/NGC4321_int_olay_dss_colour.jpg, credit: JCMT/JAC).

the whole volume of the interstellar medium. Observations of the gas can be done by studying the rotational transitions of different elements, like e.g. CO $J = 1 \rightarrow 0$ at a wavelength of 2.6 mm, or by studying UV resonance absorbtion lines of molecular hydrogen. The better part is concentrated in giant molecular clouds (GMCs). Only in these molecular clouds, the densities and temperatures are high enough for the formations of stars. With a gas fraction of 99.98% molecular hydrogen (H_2) is the most common molecule (see e.g., Omont, 2007). Molecular transitions (from FIR out to radio wavelengths, created by e.g., rotational and inversion transitions, or transitions caused by the stretching and bending of molecules) are used as observational means to study these regions. As shown in Fig. 1.7, the same astronomical object can look very different in atomic and molecular gas emission, since both gas components are tracers for very different physical properties.

1.3.2.1. The H_2 molecule

Molecular hydrogen is the most common molecule in the interstellar medium. Since H_2 has a symmetrical structure the molecule has no electric dipole moment. Molecules can only emit radiation if their dipole moment is larger than zero. For H_2 this is only possible in the first excited state (11.2 eV; T > 100 K). The corresponding emission line has a wavelength of 111 nm, in the ultraviolet waveband. Only transitions between the v and J modes of molecular hydrogen lead to the emission of radiation in the infrared and radio regime.

1.3.2.2. The CO molecule

After H_2, CO is the second most common molecule in the universe with a ratio of $n(H_2) \sim 10^4\ n(CO)$. Due to the low critical density necessary for the excitation of the CO emission ($n_{crit}(J = 1-0) \sim 740$ cm^{-3}, Forbrich 2003), occurs excitation at low densities. The excitation of the energetic lowest rotational transitions is caused primarily by the collision with H_2 molecules. Since the CO molecule has a permanent dipole moment, the de-excitation of the excited states is performed by rotational transitions between the energy levels. The excitation temperature of the lowest permitted rotational transition lies at 5.5 K. Hence the CO molecule emits radiation already at very low temperatures and the emission is radiated in the mm/radio wavelength regime. Assuming the commonly used correlation between CO and H_2 is correct, CO can be used as a tracer to indirectly observe the molecular hydrogen (e.g., Chap. 3 in Omont, 2007).

The molecular CO emission is, e.g., used to determine temperature, densities and masses of molecular clouds, molecular discs of galaxies, etc. With its strong binding energy of 11.1 eV the CO molecule is protected from further reactions with possible reacting agents from the molecules vicinity, i.e. CO is relatively stable. Furthermore, the binding energy protects the CO molecule from the influence of the UV radiation field ('self-shielding'). Isotopes of carbon monoxide used as tracers in mm astronomy are $^{12}C^{16}O$, $^{13}C^{16}O$, $^{12}C^{18}O$, $^{12}C^{17}O$ and $^{13}C^{18}O$. The most common isotope is $^{12}C^{16}O$.

Although much effort has been invested in studying the molecular gas emission in QSO host galaxies (Scoville et al., 2003; Krips et al., 2006a,b; Bertram et al., 2007), only little is known so far: Evans et al. (2001, 2006) and Scoville et al. (2003) discuss small samples of QSO host galaxies from the Palomar-Green (PG) Bright Quasar Survey. As molecular gas is the fuel for any activity process, whether it is in form of vigorous bursts of star formation (SB) or an intensively accreting supermassive black hole (in AGN), the study of its morphology and its kinematics is of key importance to understand the fueling mechanisms that keep the activity alive over cosmologically significant time scales (see e.g., Hopkins et al., 2005). The majority of nearby QSOs is associated with large reservoirs of molecular gas ($> 10^9\ M_\odot$), which militates against a quiescent nature of host galaxies. This might imply that the density of the formerly AGN obscuring medium in the central region is significantly reduced. The low-z PG QSOs and nearby Hamburg/ESO survey (HES) QSOs show CO and FIR luminosities and star formation efficiencies that are lower than it is the case for ULIRGs in the same co-moving volume. Spatially resolved data on QSOs in the local universe, however, are hardly available. Staguhn et al. (2004) present a ringlike structure with a radius of ~ 1.2 kpc in the circumnuclear molecular gas distribution of I Zw 1. For the low-luminosity QSO HE 1029–1831 (Krips et al., 2007) estimate a size of 6 ± 2 kpc for the CO source that is aligned with the optical bar. Another one, HE 2302–0857, shows an exceptionally broad CO (1−0) emission line of ~ 650 km s^{-1} (Bertram et al., 2007).

In this thesis CO observations are used to study the famous ULIRG Arp 220.

1.3.3. Maser

The term maser in short stands for *m*icrowave *a*mplificaion by *s*timulated *e*mission of *r*adiation. Maser emission from molecules requires certain physical conditions to be met. To excite the energy levels involved critical values for the density and the temperature of the molecular gas have to be surpassed: The density $n(H_2)$ has to be larger than 10^7 cm^{-3}, the temperature must be higher than ~ 400 K (e.g., Henkel et al., 2005), and the masing molecules have to undergo a population inversion. Furthermore, the velocity coherence along the line of sight through the medium has to be long enough to achieve a significant amplification of the radiation. A nearby energy source is implicit in the population inversion of the maser levels.

Other than in the lab, maser emission in the interstellar medium occurs naturally since the conditions in the ISM are typically not in thermal equilibrium and it is not uncommon to find dense gas with the right properties in the vicinities of stars (galactic masers) or active galactic nuclei. A differentiation between galactic and extragalactic maser emission has to be made due to the differences in the excitation mechanisms leading to the emission of maser radiation. Galactic masers, known since the 1960s, have been found in star formation regions in the vicinity of young stellar objects (YSOs) and in the molecular envelopes of evolved stars. Until 1977 maser emission was known only for galactic sources. The first extragalactic maser source was found in M 33 (Churchwell et al., 1977). In comparison to galactic masers, the extragalactic maser emission seems to originate from high-density molecular gas located within parsecs of the AGN of external galaxies. To date only five molecular species are known to emit radiation of maser line transitions in external galaxies: CH, H_2CO, H_2O, OH and SiO. The main work in extragalactic astrophysics is however focussed on OH and H_2O masers.

Due to the differences in the formation processes for H_2O megamasers a division into three subclasses has to be made: Disk-masers (found e.g. in NGC 4258, IC 2560), jet- or outflow-masers (observed e.g. in NGC 1052, Mrk 348) and a mixture of both, detected for example in NGC 1068, Circinus (Lo, 2005). Disk-masers are formed in a thin, warped, almost edge-on disk at radii up to only 1 pc from the galactic nucleus. Jet- or outflow-masers arise from post-shocked gas established by the impact of nuclear jets or outflows with the nucleus-surrounding molecular clouds. Furthermore, the so-called 'kilo-masers' (10^3 times more luminous than the average galactic maser, found e.g. in NGC 253, NGC 2146 Lo, 2005) are thought to arise under comparable conditions as the galactic masers. Two competing theories try to explain this weaker maser emission: Most nuclear kilo-masers are likely physically associated with star formation (M 82, Baudry & Brouillet, 1996). On the other hand it is likely that the host galaxies of kilo-masers could also harbor AGN. Hence the kilo-masers could be a weaker version of the circumnuclear disk-masers (e.g. in M 51, Hagiwara et al., 2001). To date, H$_2$O megamasers are found in 10% of observed AGNs in the local universe. They are almost exclusively found in Seyfert 2 and LINER type galaxies (Braatz et al., 1997; Kondratko et al., 2006b), i.e. mostly in spirals.

Extragalactic masers with isotropic luminosities $L_{H_2O} > 10\ L_\odot$ are classified as 'megamasers'. For comparison 'average' galactic masers have a luminosity of $10^{-4}\ L_\odot$. Megamaser emission is an excellent probe to determine the properties of nuclear accretion disks, to measure densities and temperatures in the surrounding medium of the excitation sources, to measure the strength of magnetic fields, to determine the disk stability, as well as the mass accretion rate and the mass of the nuclear engine. Additionally, megamasers are used to constrain the physical parameters of the nuclear starbursts triggered by spiral galaxy mergers, for example. The water megamaser in particular provides the only possibility to map the circumnuclear accretion disks in AGN. Additionally, the maser transition at 22 GHz is used to constrain the mass of supermassive nuclear objects. The water megamaser transition is a very valuable asset for obtaining the exact distances to external galaxies as well as for studying interactions between nuclear jets and dense molecular material ('jet-masers').

The work on masers in this thesis is focused on water megamasers in nearby QSO host galaxies.

2. Observations in the radio and mm wavelength regime

In radio astronomy, paraboloid reflectors in the forms of dishes or horn antennae, are used for scientific purposes. Even a simple dipole can be used as an antenna. Observations in the mm and radio wavelength regime are conducted with the help of heterodyne receivers, for observations of line emission, bolometers, for observations of the continuum, or Schottky diodes. Heterodyne receivers use the frequency mixing technique: The frequency of the astronomical signal is mixed with the signal of a so-called local oscillator (LO) and is thereby shifted to a lower frequency, which is easier to handle for the electronics, since there is a lack of amplifiers adequate for high frequencies. The transition to observations in the FIR regime is characterized by the use of so-called bolometers. Instead of executing spectral decomposition, bolometers directly detect the received signal at the observed frequency with a very large bandwidth and are therefore used for observations of the continuum broadband emission.

Telescopes are divided into two subclasses: Single-dish telescopes and interferometers. Freely steerable single-dish telescopes have diameters up to ~ 100 m (e.g., the Effelsberg and the Green Bank telescopes). But here the physical feasibility already reaches the limit. Due to gravitational forces, dishes with larger diameters are not possible to build with the surface accuracy needed. But even this huge telescopes are not large enough to study many objects of interest in the resolution needed. This is due to the fact, that the angular resolution of a telescope follows the formula *resolution* $\propto \frac{\lambda}{D}$. This means that larger telescopes (large diameter D) have a better angular resolution, but the longer wavelengths (λ), that are studied in radio astronomy, decrease the spatial and angular resolution significantly. This poses problems especially for studies of objects that aim at great detail and for sources at higher redshifts, considering that the resolution also scales with the distance to and the size of the object. In order to build bigger telescopes without having to deal with the construction problems of very large dishes, several small single-dish telescopes are connected to each other to synthesize one very large telescope. Interferometers provide the desired angular resolutions (down to several micro-arcseconds) but have the disadvantage of longer integration times in order to achieve the desired signal-to-noise ratios.

In the following, the basic principles of single-dish telescopes and interferometers, along with the telescopes used for the observations (Effelsberg 100-m telescope, VLA, PdBI), are shortly described. Additionally, techniques used for the data acquisition are presented.

Specific details on every observational data set studied are given in the corresponding chapters of this thesis.

2.1. Single-dish telescopes

A single-dish telescope (Figs. 2.2, 2.1) is mathematically described by its power pattern $P(\nu,\theta,\phi)$. $P(\nu,\theta,\phi)$ characterizes the directional dependence of received or emitted signals. The model describes an antenna as a directionally independent emitting dipole. In reality the energy emission is dependent on the solid angle:

$$P(\nu,\theta,\phi) = G(\nu,\theta,\phi)P \qquad (2.1)$$

G is the so-called gain or directivity of the telescope at a certain frequency ν. For distinct directions the antenna receives a much higher signal intensity than for other solid angles. The beam solid angle of an antenna is described by :

$$\Omega_A = \int_0^{2\pi}\int_0^{\pi} P_n(\nu,\theta,\phi) \sin\theta \, d\theta d\phi \qquad (2.2)$$

2.1. Single-dish telescopes

An ideal antenna would have a distribution of $P_n = \sin^2 \theta = 1$ within a small solid angle. Out of this range all power would be zero ($P_n = 0$). Unfortunately, real antennae show solid angles in which, for certain areas of θ and ϕ, much more power is received in those areas than out of them. This area is termed the main lobe or beam Ω_M:

$$\Omega_M = \iint_{beam} P_n(\nu, \theta, \phi) d\Omega \qquad (2.3)$$

For real antennae the intensity of the signals in the side lobes (those areas outside the beam where less signal is received) is not negligible. Interferences (so-called radio frequency interferences or RFI) of e.g. earthbound radio transmitters or satellites are detected in those areas of the power pattern. Additionally strong astronomical radio sources in the vicinity of the observed object contribute to this effect. In this way the 'sources' in the side lobes (either of artificial or astronomical origin) influence the quality of the directional measurements. This quality is described by the main beam efficiency η_M:

$$\eta_M = \iint \frac{\Omega_M}{\Omega_A} \qquad (2.4)$$

η_M characterizes which part of the power pattern is situated within the main lobe or beam. The beam is usually characterized either by the half power beam width (HPBW) or the full width at half maximum (FWHM) of the maximum intensity of the received signal.

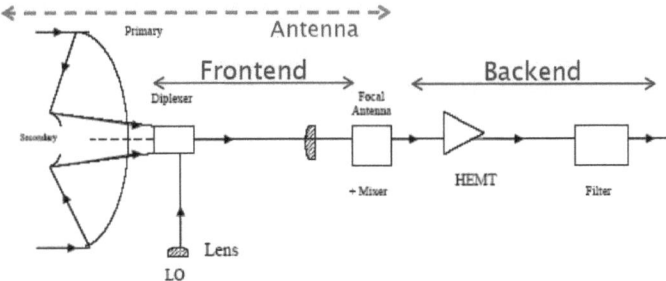

Figure 2.1.: A simplified schematic diagram of a single-dish telescope (taken from the 5th IRAM 30 m Summer School, 2009).

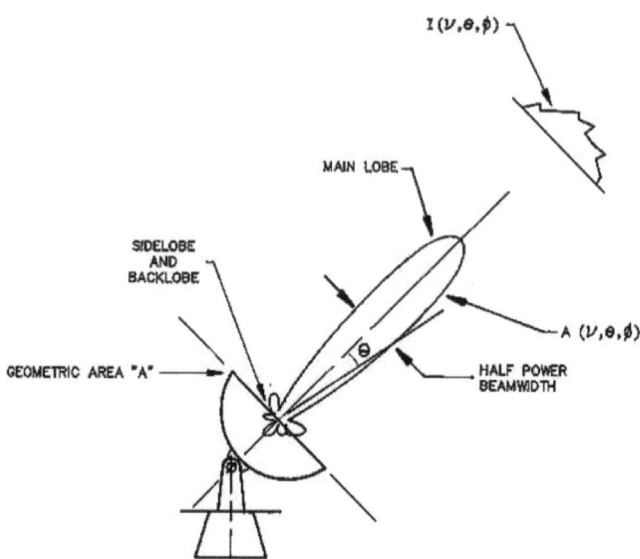

Figure 2.2.: A schematic view on a single-dish telescope and its main characteristics: lobes, HPBW (taken from the Eleventh Synthesis Imaging Workshop, Socorro, 2008).

2.1.1. The Effelsberg 100-m telescope

Built in 1971, with a diameter of 100 m, the Effelsberg radio telescope (Fig. 2.3) is the biggest fully steerable single-dish telescope in Europe. It is located to the southwest of the city of Bonn, Germany, from where the telescope is operated by the MPIfR. In conjunction to the paraboloid primary mirror, a 6.5 m secondary mirror is installed, providing the observer with a second set of choices for receivers. Observations are possible in a frequency range from 400 MHz (73 cm) to 96 GHz (3 mm). Angular resolutions between 11′ (at 800 MHz) and 10″ (at 96 GHz) are achievable. Astronomical objects can be observed in spectral line mode as well as in the continuum. All single-dish data obtained in the framework of the thesis were obtained with this instrument.

Figure 2.3.: The Effelsberg 100-m telescope by day and night.

2.2. Interferometry and interferometers

Interferometry is done by the combination of several single-dish telescopes. They can be placed in close proximity with baselines on the order of meters or kilometers, but they can also be spread over longer distances, even continents (Very long baseline interferometer, VLBI). For interferometers the length of the telescope separation (baseline) determines the angular resolution in the resulting image, i.e. the larger the baselines are, the smaller the structures get which can be imaged. On the other hand, small telescope spacings make it possible to detect the extended emission. A simplified diagram of an interferometer is shown in Fig. 2.4. The received signals from the respective single telescopes cannot be combined simply by adding the signals together. In order to receive useable data from the whole array the signals from the single telescopes have to be coherent, i.e. one has to make sure that the radiation emitted by the observed object at one time, which is received by the single telescopes with slightly different time delays is combined in the right way. In order to preserve the coherence an artificial time delay between the telescopes of the interferometer array has to be introduced. For this reason every telescope of the array has its own high-precision clock. As the separations

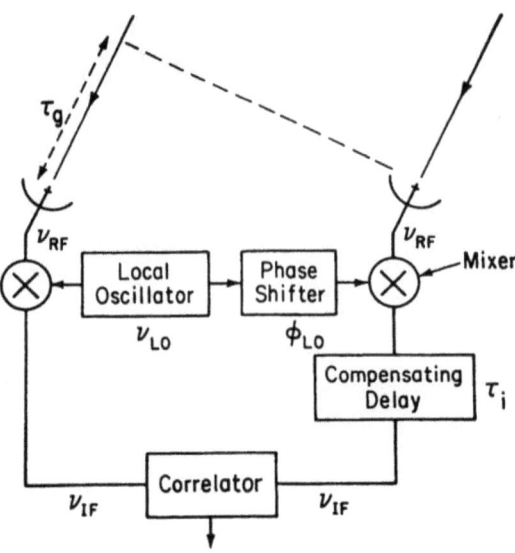

Figure 2.4.: The simplified schematic diagram of a two-element interferometer, including the instrumental time delay, to compensate for τ_g, and the frequency conversion to the intermediate frequency v_{IF} (taken from Taylor et al., 1999).

between the single telescopes are known, the needed artificial time delay can be derived and incorporated into the signal combining process: The signal is received by the individual telescope dishes. Afterwards, the frequency of the incoming signal is conversed by a local oscillator (LO) because it is technically convenient to perform the following operations at an intermediate frequency. Later the signals are amplified, filtered, and the artificial delay is introduced before all the signals are combined in the correlator.

In contrast to single-dish telescopes interferometers do not measure the intensity distribution of the observed object $I(x,y)$ directly but its fourier transform $V(u,v)$. x and y are the coordinates in the plane on the sky and u and v are the coordinates in the fourier space, i.e. the coordinates of the projected baseline for each of the single telescope pairs seen from the source. The intensity distribution of the source can be obtained by an inverse Fourier transform:

$$I(x,y) = \int\int e^{-2\pi i [xu+yv]} V(u,v)dudv \qquad (2.5)$$

In practice, due to the limited observing time, the visibility function $V(u,v)$ is not known in the whole uv-plane but is sampled discretely. The intensity distribution determined by the Fourier transform of the discretely sampled V, however does not represent the true structure of the object yet. This intensity distribution, which is still convolved with the so-called dirty beam B of the interferometer, is termed the dirty image. As the result of the de-convolution of the dirty image with the dirty beam, the true intensity distribution of the observed source is obtained.

2.2.1. The Very Large Array (VLA)

The Very Large Array (VLA, Fig. 2.5) is an interferometer consisting of 27 single-dish telescopes located on the Plains of San Agustin (2 124 m above sea level), fifty miles west of Socorro, New Mexico in the US. The radio telescopes are arranged in a Y-shaped pattern and can be grouped in four different array configurations (A, B, C and D). Each antenna dish has a diameter of 25 m. Antenna separations range between 1 km up to 36 km in the A array. The antennaes are outfitted with receivers for frequencies between 73 MHz and 50 GHz, for both continuum and line observations, and are therefore able to probe the meter to millimeter wavelength regime. Depending on the array configuration and observation band (frequency), angular resolutions between $24''$ and $0.05''$ can be achieved.

Sources from a subsample of the Cologne nearby QSO sample have been studied with the VLA in 21 cm HI line emission.

2. Observations in the radio and mm wavelength regime

Figure 2.5.: The Very Large Array (VLA).

2.2.2. The IRAM Plateau de Bure Interferometer

This interferometer consists of a group of six antennae. It was build starting in 1985 on the Plateau de Bure (2 550 m above sea level) in the French Alps approximately 100 km from Grenoble, France, where its operating institute, IRAM, is residing. Each of the six 15 m antennae is equipped with high-sensitivity receivers covering a frequency range from 80 to 370 GHz (3 to 0.8 mm). The telescopes are arranged in T-shaped configuration on a north-south and an east-west track. The longest possible baseline extends out to 760 m in the A array in the E-W direction, corresponding to a single-dish telescope of 760 m in diameter. Angular resolutions of $8''$ down to $0.5''$ depending on array configuration and frequency setup, can be achieved.
All millimeter CO observations of Arp 220 have been carried out with this interferometer.

2.3. Observation methods

Since interferometers or interferometric arrays are made-up by single-dish telescopes, both 'instruments' can achieve very different resolutions and sensitivities, but they still have, in principle, the same properties. Handling an array of telescopes, in comparison to only one single instrument, increases the effort by several orders of magnitude. Each telescope has to be equipped with the corresponding technical facilities (receivers, backends, and so on). Also, the technical complexity of the data handling and the amount of data obtained, grows enormously with increasing number of telescopes. In interferometric observations each telescope has a number of baselines with the other telescopes from the array producing interference fringes and therefore deliver valuable data. Furthermore, interferometers can be used for observations with different angular resolutions, since they can be arranged in different patterns. Hence, they are more flexible in terms of requisitions to the observation of very different astronomical objects.

Although all of this seems to point towards completely different handling modes of observations, interferometers still have the same properties of single-dish telescopes. Hence, observation techniques developed for single-dish telescopes are also used for interferometric observations.

In the following, a brief description of the observing techniques applied during the single-dish and interferometric observations is given.

2.3.1. Position-switching

For this observing technique two positions on the sky are necessary: The position of the object to be observed (ON-position) and a position, the so-called OFF-position, which is used as a reference and has to be free of emission in the observed frequency range. In order to reduce differences in the optical depth of the air masses, OFF-positions usually only differ in their azimuth relative to the ON-position. As the optical depth of atmospherical layers depends on the elevation, it is useful to choose angular distances between ON and OFF to be as small as possible, always under the condition that the OFF-positions stays emission-free. Observations consisting of several pairs of short ON-OFF integrations improve the baselines in the spectra significantly compared to one long ON-OFF scan. This approach is especially convenient to remove interfering signals coming from the receivers or the atmosphere. Therefore the ON and OFF spectra are subtracted from each other. Objects observed with this method show improved spectral baselines and a diminished influence of atmospheric disturbances. However, one downside of this method is the huge amount of time spent by slewing the telescope back and forth between the ON and OFF-positions.

Position-switching is used preferably with objects that have large dimensions and exhibit complex spectra. Furthermore spectra with broad line wings as well as spectra with hyperfine structure are studied with this method. Another application of this technique is the analysis of spectra with several velocity components spread over large ranges in the velocity domain.

In this work the position-switching technique was applied to all 21 cm HI line emission observations.

2.3.2. Load-switching

For the load-switching method no reference position is required. This is a big advantage compared to position-switching technique, the problem to find an emission free reference position is avoided and time to slew the telescope is significantly decreased, which leads to an increase of the on-source integration time. Instead of switching between the positions, here the frontend (receiver) continually switches between the signal from the line-source and a cold load at a constant temperature. The telescope is only tracking the position of the observed object on the sky. After the subtraction of both measurements the resulting spectrum shows two emission lines of different rest frame velocity and polarization. To obtain a spectrum appropriate for analysis, a convolution of the two emission lines has to be carried out.

With the load-switching technique, one highly efficient observing method resulting in an excellent signal-to-noise ratio (SNR), is available to the community. Load-switching is not suited for observations of faint or broad emission lines. Whereas for narrow emission lines without fine or hyperfine structure and with only a single velocity component this technique is very much appropriate. For very extended source objects load-switching is a very good alternative to position-switching since the problem of finding emission-free reference positions does not exist. One disadvantage of load-switching, however, is the often bad quality of the spectral baselines. They can show 'baseline ripples', that are produced by 'spillover' emission from the side lobes of the telescope or by very strong continuum sources.

This observing technique was used in the search of the Cologne-QSO sample for H_2O maser emission only.

3. The Arp 220 merger on kpc scales

3.1. Introduction

Ultraluminous IR galaxies (ULIRGs) are a galaxy class characterized by an IR luminosity of $10^{12} L_\odot \leq L_{bol} \sim L_{IR} [8-1\,000\,\mu m] \leq 10^{13} L_\odot$. They are thought to be powered by either a strong nuclear starburst (Joseph, 1999), a heavily obscured active galactic nucleus (AGN, Sanders, 1999) or a mixture of the two. Whatever the power source may be, a large molecular gas concentration in the central kiloparsecs is needed in order to sustain it (see e.g., Sanders et al., 1988). Various scenarios linking the evolution of ULIRGs to QSOs have also been suggested (e.g. Sanders et al., 1988).

At a luminosity distance of about $D_L = 78$ Mpc ($\Omega_\Lambda = 0.7$, $\Omega_M = 0.3$ and $H_0 = 70$ km s^{-1} Mpc^{-1}) and with a luminosity of $L = 1.4 \times 10^{12} L_\odot$ (Soifer et al., 1987) Arp 220 is a prototype ULIRG. Extended tidal tails, apparent in the optical (Arp, 1966), as well as the double nuclei revealed in the radio and the near-infrared (NIR) regime (e.g. Norris, 1988), indicate that the galaxy is in the final stage of merger. The evolutionary scenarios linking ULIRGs to QSOs allow the possibility that a weak, but growing, AGN (Taniguchi et al., 1999) may lie at the center of Arp 220. Competing theories claim a pure nuclear starburst contribution, contributions from hot cores or a mixture of AGN and starburst to explain the high IR luminosity (Sakamoto et al., 2008), arguing with the rather high star formation rate of 340 $M_\odot$yr^{-1} (Baan, 2007). Fe Kα emission at 6.3 keV was detected at slightly more than 3σ significance, although there is no angular resolution for resolving the galaxy. This may also indicate the presence of an AGN deeply embedded in dust.

The overall gas content was estimated to be $\sim 9 \times 10^9 M_\odot$ (Scoville, 2000). The fact that [OIII]λ5007, which is a good AGN tracer but also sensitive to star formation, is not detected in the optical supports the extremely high extinction. Sakamoto et al. (1999) found that the extinction towards the main components of Arp 220 can be extremely high. They derived an average extinction perpendicular to the nuclear disks of $A_V \sim 1\,000$ mag, based on the mean gas and dust surface densities from CO(2−1) and continuum observations. Optical observations reveal that the galaxy is clearly crossed by a dust lane spreading from NE to SW (see Fig. 3.1a). High resolution radio and near-infrared imaging revealed a complex nuclear structure: The core of Arp 220 appears to have (at least) two different nuclei (or nuclear regions, Fig. 3.1). They are identified as western and eastern nuclei, separated by about 0.4 kpc (at this redshift 0.018 kpc correspond to about 1″). The eastern nucleus is divided into two components: North-east (NE) and south-east (SE, Downes & Eckart, 2007). CO(1−0) observations revealed an underlying rotating kpc-sized molecular gas disk (Scoville et al., 1997). The presence of compact sources identified as supernovae (at 13, 6, and 3.6 cm) was detected by Parra et al. (2007). A comparison between CO and near-IR data shows consistency in the number of sources, but not in their relative position. This may be due to: (1) the fact that different wavelengths trace different gas phases, or (2) differential extinction.

Following it's position in the proposed evolutionary scenarios (see earlier in the text), Arp 220 can provide the link between ULIRGs and QSOs. Furthermore it can be considered as a benchmark in understanding these complex objects in detail. Arp 220 may also turn out to be useful to explain the rotation-like structures detected in high-redshift ULIRGs (see Förster Schreiber et al., 2006). Studying Arp 220 and other ULIRGs in the local universe, using high spatial and spectral resolution observations, may allow to draw conclusions for radio detected galaxies at high redshifts, which cannot be studied in this great detail, yet.

3.2. Observations and data reduction

The HST/WFPC2 F814W archive image was reduced 'on the fly' using the best reference files at the moment of retrieval. In this chapter this image is shown only as a reference. Details on the NICMOS data reduction can be found in Scoville et al. (1998). CO 1 and 3 mm line and continuum observations with different spatial scales and resolutions were carried out with the IRAM Plateau de Bure interferometer (PdBI) in different configurations during the years 1994, 1996, 1997 and 2006, respectively. Depending on the observational setups and the data reduction, the synthesized beam sizes range from $0.30'' \times 0.30''$ to $4.99'' \times 3.50''$. The CO(1−0) high (Downes & Solomon, 1998) and low resolution data were taken in winter 1996 with a bandwidth of 500 MHz and a channel separation of 2.5 MHz, resulting in a velocity range of $\sim 1\,300$ km s^{-1} and a velocity channel width of 7 km s^{-1}. Using natural (high resolution) and uniform weighting (low resolution) in the mapping procedure gave beam widths of $1.58'' \times 1,11''$ (high resolution) and $4.99'' \times 3.5''$ (low resolution). The resulting CO data cubes were analyzed with the MAPPING package of IRAM's GILDAS[2] software.

3.3. Unveiling the central kiloparsecs

With CO PdBI observations and archival HST images, the dust distribution and the distribution of the star forming regions can be analyzed, the position of the supposedly true nuclei can be disentangled, the kinematics can be studied on different spatial scales, and the data make it possible to draw an interesting picture of the central arcseconds ($''$) of Arp 220.

---

[2] http://www.iram.fr/IRAMFR/GILDAS

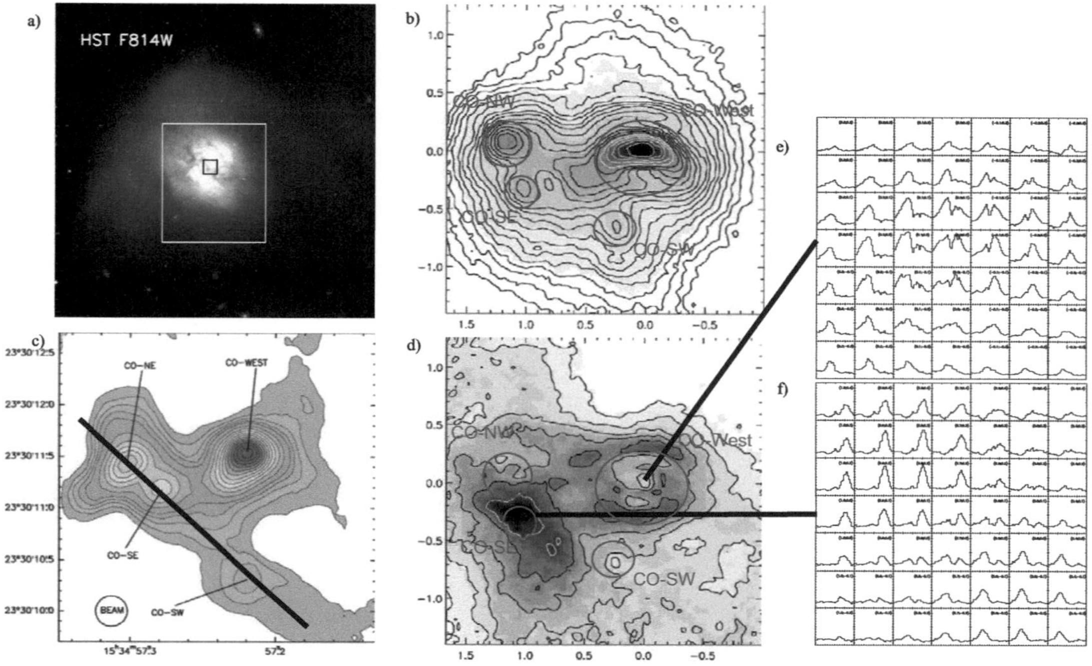

3. The Arp 220 merger on kpc scales

Figure 3.1.: a): HST WFPC2/F814W I-band image of Arp 220 (North is up, east to the left). The FoV (field of view) of the CO (2−1) (black rectangle) and the CO (1−0) maps (white rectangle) are indicated. b): HST 2.22 µm NIR contour and grey-scale image of the central 2.5″ of Arp 220. Coordinates are offsets in α and δ from the 2.2 µm peak (Scoville et al., 1998). c): Central 3″ of Arp 220 in CO (2−1) emission (Downes & Eckart, 2007). The line indicates the direction of the position-velocity cut through the eastern nucleus (see Fig. 3.2a).d): HST 2.2 µm/1.1 µm color map representing the dust distribution in Arp 220 (Scoville et al., 1998). CO (2−1) spectra across Arp 220-West (e) and Arp 220-East (f).

3.3.1. Arp 220-West

The western nucleus of Arp 220 is the brightest source in the near-infrared ($F_\nu = 20$ mJy$''^{-2}$, Fig. 3.1b), in the high resolution CO(2−1) map (peak flux density $F_\nu = 56.5$ Jy km s^{-1}, Fig. 3.1c) and in the 1.3 mm continuum ($F_\nu = 79$ mJy beam^{-1}; Downes & Eckart, 2007). The 2.2 μm/1.1 μm color map (peak value 15.8, Fig. 3.1d) however reveals that the peak of the dust distribution is not associated with this nucleus, but ~ 360 pc away on the position of CO-SE. The CO(2−1) spectra of Downes & Eckart (2007) (Fig. 3.1e) show deep absorption that can be explained by the presence of cooler gas in the foreground of the emission peaks. They also found that the complex CO emission can be modeled best by a cooler (50 K) molecular gas ring or disk around a compact, hotter (170 K) dust core. In their 1.3 mm continuum data these authors found a size for the dust source of $0.19'' \times 0.13''$, which is very small. Hence the black body luminosity of this compact dust source ($\sim 10^{12} L_\odot$) implies that the energetic process powering Arp 220-West cannot be of starbursting nature. For the CO(2−1) emission centered on the dust source Downes & Eckart (2007) furthermore discovered a steep velocity rise towards the western nucleus. This behavior is a further indicator for the presence of a black hole in this nucleus. Hence, Downes & Eckart (2007) concluded that size, luminosity and rotation pattern observed in their CO(2−1) data support the presence of a black hole in Arp 220-West.
In the simulations of the merger in Arp 220 (see Sect. 3.6), the western nucleus is the core of merger component I.

3.3.2. Arp 220-East

The eastern nucleus, which is veiled at optical wavelengths, can be resolved into a NE and a SE part (see Fig. 3.1c). Former publications (Downes & Solomon, 1998; Sakamoto et al., 1999), before Downes & Eckart (2007), did not resolve the eastern nucleus in two components, in CO emission. The velocity gradient which they used to perform their modeling was the one shown

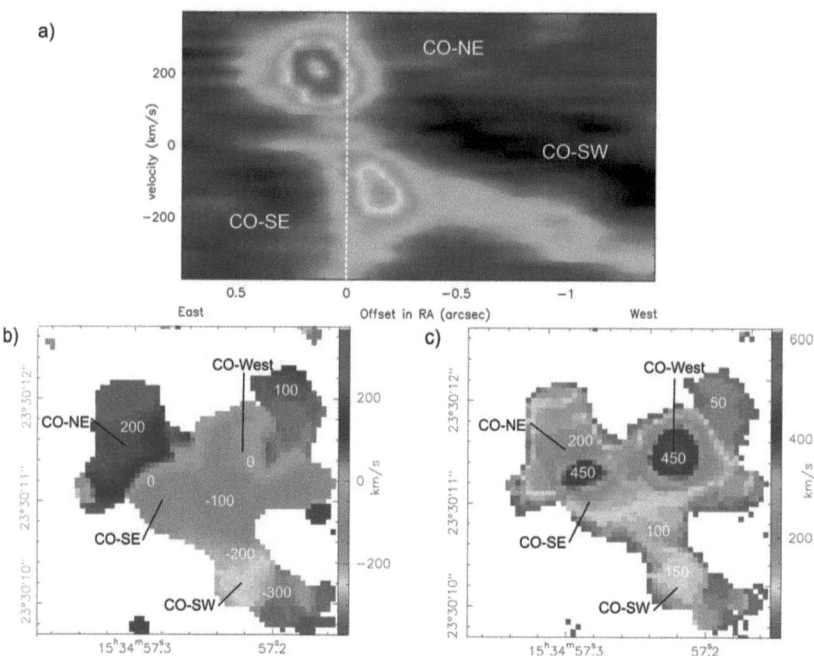

Figure 3.2.: a) East-west CO(2−1) position-velocity cut through Arp 220 from CO-NE to CO-SW, with the 1.3mm continuum subtracted, considering CO-SE as the rotating center and a radius of about 0.3″. b) CO(2−1) velocity field of the central 3″ of Arp 220, with a beam of 0.30″. c) CO(2−1) velocity dispersion distribution, showing two prominent peaks at the positions of CO-West and CO-SE and a possibly fainter third one towards CO-SW. Representative values are indicated in white.

in Fig. 3.2 – in this thesis for the first time the CO intensity maps are actually confirmed by the p-v diagram presented in this figure (Fig. 3.2a). Arp 220-East is the secondary source in the distributions of near-IR, CO(2−1) and CO(1−0) emission. With the higher resolution CO observations a more detailed comparison to the near-infrared observations of Scoville et al. (1997) was possible: The maximum of the near-IR color map (Fig. 3.1d) is located at the position of CO-SE and hence points at CO-SE as the peak of the dust distribution. Using $H-K$ and a standard extinction law (Rieke & Lebofsky, 1985), the A_V for the SE and NE nuclei in the NIR are 24 and 18 mag, respectively. In addition, the CO(2−1) spectra show the deepest absorption in the CO-SE region (Fig. 3.1f). Along with the velocity dispersion distribution (Fig. 3.2c), peaking on the position of CO-SE rather than on CO-NE, this seems to point out CO-SE as the choice to be identified with a deeply embedded and/or highly dust obscured nucleus of Arp 220-East. The steep rise in the width of the emission lines is about 200–250 km s^{-1} compared to the region in-between the eastern and the western nuclei of Arp 220. Continuing in the direction CO-West, the line width then rises again steeply to values of ~ 400–500 km s^{-1}.

With the information provided by the emission line width peak located on CO-SE (Fig. 3.2c) and the structure of the velocity fields (Figs. 3.2b, 3.3), it is possible to assert the presence of rotating material around the SE nucleus. A possible third nucleus, CO-SW (south-west), is clearly identified in the near-IR (Fig. 3.1b), and in CO (2−1) (Fig. 3.1c, line width enhancement ~ 50 km s^{-1} compared to the surroundings of CO-SW), although it is not detected in the 1.3 mm continuum (see Fig. 2 of Downes & Eckart, 2007). In fact it approximately coincides with the secondary peak of the 3–7 keV map. The extinction derived, from the NIR, is $A_V = 17$ mag. The structure of the line width map (Fig. 3.2c) suggests a further mass peak, towards CO-SW. This enhancement could be produced by a mass concentration, but it can also be produced by turbulent motions due to the merger process. Although the true nature of CO-SW is still unclear, it could possibly host an additional very faint nucleus of a possible minor merger component (mergers that involve a gas-rich disk galaxy and a bound companion or satellite that usually has $\lesssim 10\%$ of the mass of the gas-rich galaxy), a huge star forming region, or a remnant of the merging process from one of the colliding galaxies.

In the simulations of the merger in Arp 220 (see Sect. 3.6), the CO-SE and NE peaks are handled as the core of merger component II.

3.4. The different kinematic scales in Arp 220

All velocity fields derived from the nuclear regions show rotation-like patterns at different scales (see Figs. 3.2b, 3.3). Only in the outer regions, observed at optical wavelengths, the velocity pattern is disordered and apparently dominated by the merger (Colina et al., 2004). In the inner 2 kpc there

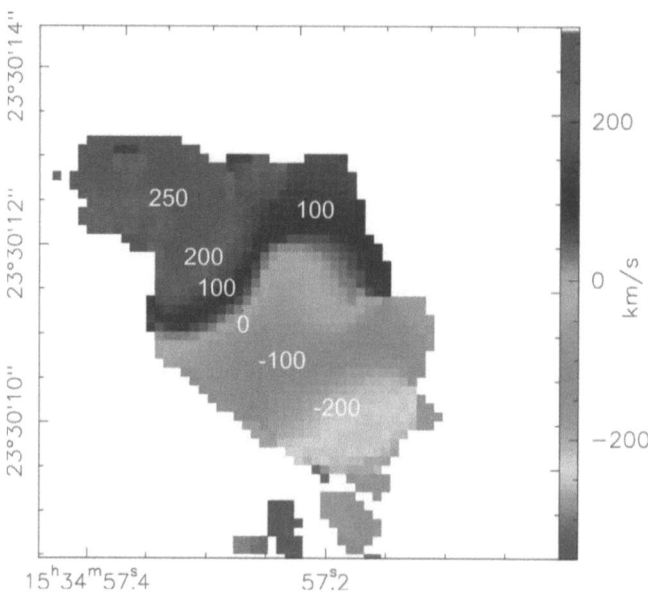

Figure 3.3.: Velocity field of the high-resolution CO (1–0) emission observations of the central 6″ of Arp 220, with a beam size of 0.60″. Representative values are indicated in white.

are influences from the outflows coming from the dust-enshrouded nucleus on the ionized gas (Arribas et al., 2001).

A CO (2−1) position-velocity diagram obtained for Arp 220-West (Downes & Eckart, 2007) shows strong variation from negative to positive velocities over the central 0.2″. For Arp 220-East (Fig. 3.2a) the p-v diagram also shows a variation from negative ($v \sim 200$ km s^{-1}) to positive velocities ($v \sim 100$–150 km s^{-1}), considering CO-SE as the rotating center and a radius of about 0.3″. The rotational motion found for CO (1−0) (Fig. 3.3) corresponds to the overall rotation of the underlying molecular disk (total H$_2$ mass $M_{H_2} = 2.9 \times 10^{10}$ $M_\odot$) discovered by Scoville et al. (1997). Here the velocity gradient spans a range of ~ 400–500 km s^{-1}. The rotational pattern seems to have its center at the position of CO-SE. The velocity filed for CO (2−1) however, looks different (Fig. 3.2b): Since the CO (2−1) line transition probes higher gas densities, one can look deeper into the gaseous environment with this higher molecular transition. Therefore, a more complex rotational pattern emerges. It seems that there are indications for rotation in CO-West, CO-NE and SE, with the center located in-between the two intensity peaks, and in

CO-SW. The velocities differ on scales of about $\Delta v \sim 200$ km s^{-1} (CO-West), $\Delta v \sim 400$ km s^{-1} (CO-NE & SE), and $\Delta v \sim 200$ km s^{-1} (CO-SW). These rotational motions in the molecular gas of Arp 220 was modeled as e.g., one underlying molecular disk (Scoville et al., 1997) with additional gaseous disks in the eastern and western nuclei (Sakamoto et al., 1999), and explaining the flux concentrations in Arp 220-West and East by crowding of inclined circular orbits of a warped disk (Eckart & Downes, 2001). Downes & Eckart (2007) modeled high-resolution CO(2−1) observations for Arp 220-West. They found indications that the western nucleus, in fact, contains a black hole with an accretion disk.

In addition to the emission line width peaking in the SE nucleus (with about the same value as in the Arp 220-West) also the maximum of the $H-K$ color map is located in CO-SE. Based on the velocity field, it is possible to assert that there is material rotating around it.

On the position of CO-SW the velocity field shows a rotation pattern centered in this region and the p-v diagram shows a local gradient that may be interpreted as rotation, too (Fig. 3.2).

3.5. The large scale picture

The internal structure of Arp 220 has been studied extensively over the years. Nonetheless, little effort has been dedicated to study the more external regions although there are indications for activity in these regions, too.

The elliptical-like envelope (see e.g., the CO(1−0) emission in Fig. 3.3) may be consistent with a late merger phase within the elliptical-through-merger scenario. The fact that near-IR light profiles follow an $r^{1/4}$-law (Wright et al., 1990) corroborates the last statement, because this law found by de Vaucouleurs (1953) describes the characteristic dependance, of the surface brightness distribution of the galaxy depending on the radius, of elliptical galaxies.

3. The Arp 220 merger on kpc scales

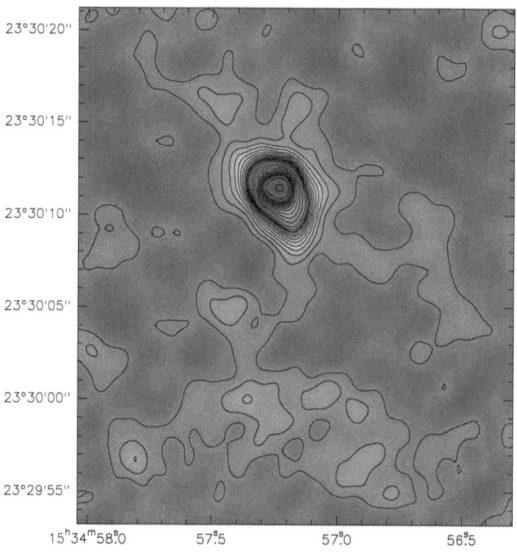

Figure 3.4.: Integrated intensity map of the CO(1−0) emission in Arp 220 integrated over a velocity range of 730 km s^{-1} with a beam size of 1.57″ × 1.11″ and with contours starting at the 1σ level, with increasing intensity in steps of 2σ.

Until today it is not clear yet what happens in the outer regions of this object. For the first time it was tried to detect and study the molecular gas outside the central 3″. There are indications, in the CO(1−0) low resolution maps, for emission about 10″ towards the south, as well as to the north and to the west of the nuclei (Fig. 3.4). The preliminary velocity field of the large scale structure coincides fairly well with the IFU (integral field unit) optical data from Colina et al. (2004). A recent more careful re-analysis of the data used in Downes & Solomon (1998) makes me confident, that the detection is real. A further comparison with other low-resolution CO observations yielded the same result.

Furthermore, the molecular mass for the extended emission towards the south of Arp 220 was determined by:

$$M = \langle N \left[\frac{H_2}{X}\right] \rangle \cdot A \cdot 2 \cdot \mu \cdot m_H \quad M_\odot \; . \tag{3.1}$$

M is the total mass of molecular hydrogen of the extended emission region, $\langle N(H_2) \rangle$ is the hydrogen column density determined from the low-resolution

CO$(1-0)$ data, A is the area projected onto the sky in the distance D_L in cm^{-2}, m_H is the mass of one hydrogen atom (1.67×10^{-24} kg) and μ is the average molecular weight per hydrogen atom ($\mu = 1.36$). Applying this formula to the CO data for this region results in a total molecular gas mass (within the 1σ level contour line) of $\sim 1 \times 10^8 \, M_\odot$.

3.6. Identikit simulations

3.6.1. The Identikit model

The Identikit model (Barnes & Hibbard, 2009) is a tool to simulate mergers of galactic disks with the help of test particles. Test particles (representing stars and clumps in this simulations) are used due to their ability to reproduce features such as bridges, tails and shells, in particular, very well. The simulations described by the Identikit model are based on collisionless N-body simulations (for the validity of the assumption of collisionless particles see Sect. 3.6.3). The simulations presented here are based on a model where a bulge, a disk, and a halo component are included.

Fifteen parameters are used to describe the model. In Fig. 3.5 a representation of the parameter space used to fit the data is displayed in a cylindrical coordinate system. The initial orientation of the two galaxy disks related to the orbit is described by the inclination i and the pericentric argument ω of each galaxy disk. Parameters defining the orbit are the eccentricity of the orbit e, the pericentric separation p and the mass ratio μ of the two disks. In the Identikit model the eccentricity is fixed at a value of one. The evolution time t describes the age of the merger in internal units of $-2 \leq t \leq 8$. $t=0$ represents the point in the merger evolution at which, in the configuration of the merging/interacting system, the idealized Keplerian orbit reaches the pericenter. Size and form of the parameter space are determined by the error of each simulation parameter (shaded area in Fig. 3.5). The shaded area with the point in the center (Fig. 3.5) represents the approximate location of the parameter space where the data used here show the best fit.

3. The Arp 220 merger on kpc scales

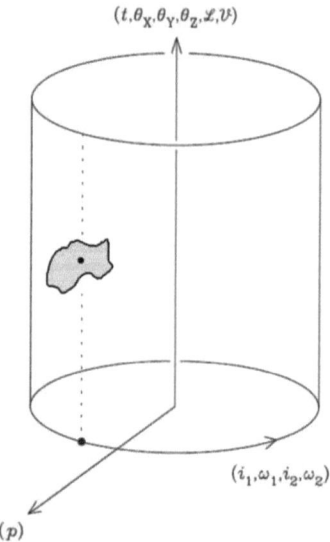

Figure 3.5.: A presentation of the parameter space used in the Identikit model simulations. The radial coordinate represents the initial orbit described by p, the azimuthal coordinate represents the disk orientations (inclination i and pericentric argument ω of each galaxy disk), and the vertical coordinate represents the parameters chosen after the simulation is run (evolution time t, the viewing angles Θ_X, Θ_Y, Θ_Z). A conventional N-body simulation explores the parameter subspace represented by the dotted line, while a single Identikit simulation can explore the entire cylindrical surface. Adapted from Barnes & Hibbard (2009).

3.6.2. Simulation results

Optical HST archive composite image[3] and mm-CO (obtained by Eckart & Downes, 2001; Scoville et al., 1997) data are used to fit the Identikit model simulations. The HST images were used to identify the 'overall shape' (x-y plane) and the CO data were used to determine a measure on the position-velocity diagrams (x-v and v-y plane).

In order to obtain a grid of possible models and to obtain the 'best model fit' from that grid, many different simulations were run systematically. To get a first estimate on the model parameters for Arp 220, the model of the Antennae galaxy was used as a starting point, especially for the inclinations i and pericentric arguments ω of the galaxy disks (Antennae: $i_1 = i_2 = 60°$, $\omega_1 = \omega_2 = -30°$, Toomre & Toomre, 1972; Barnes, 1988). The Antennae galaxy has a

---

[3] from http://imgsrc.hubblesite.org/hu/db/images/hs-2008-16-aq-full_jpg.jpg (Credit: NASA, ESA, the Hubble Heritage (STScI/AURA)-ESA/Hubble collaboration, and A. Evans (University of Virginia, Charlottesville/NRAO/Stony Brook University))

3.6. Identikit simulations

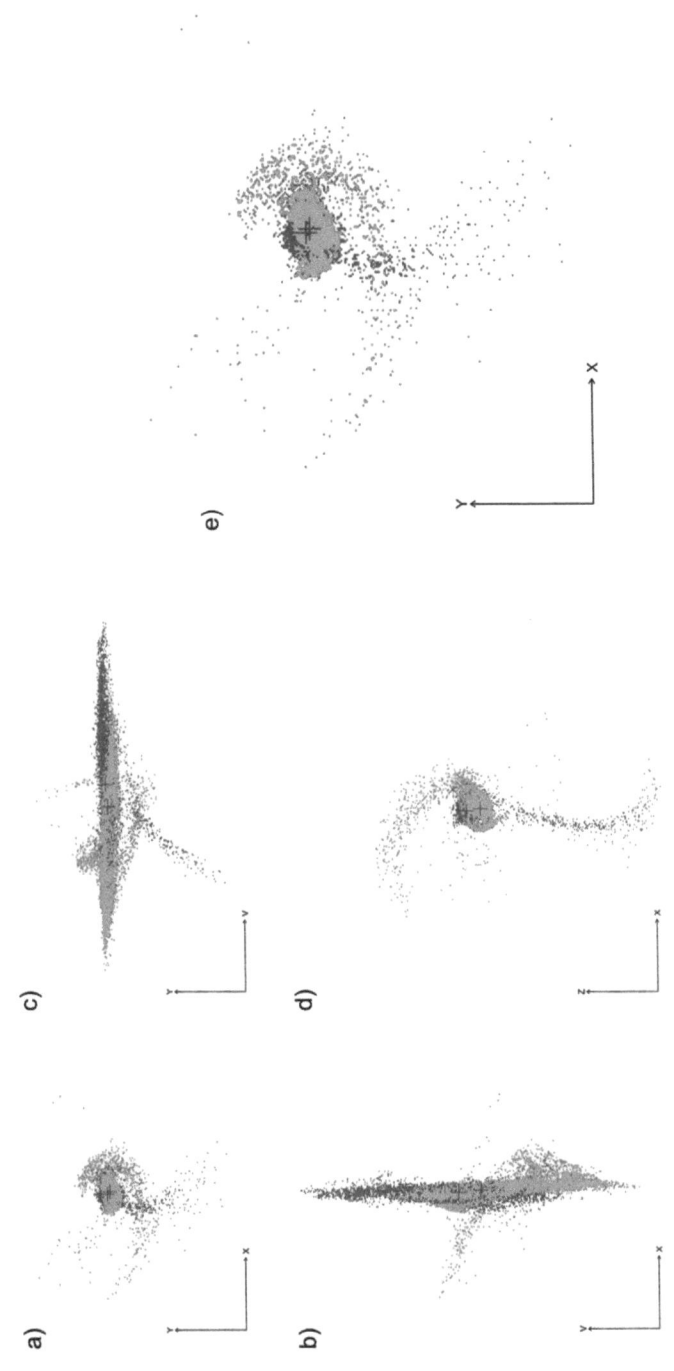

Figure 3.6.: Identikit results for the model with the best fit to the available data. a) to d) show the model projected onto the x-y (a), x-v (b), v-y (c) and x-z (d) planes. In e) the magnified x-y plane is shown.

similar morphology, compared to Arp 220, and it also shows tidal tails from both merging galaxies, but the viewing angle is different. The eccentricity was fixed internally from the start at a value of $e = 1$. In the Identikit simulation tool, only two values for the mass ratio μ were available to choose from, 1:1 and 1:2. Actually, the simulations were performed for both mass ratios. At the start of the simulations all possible values for the pericentric separation p, in the combination with the merger time scale t, were run. After the first few crudely determined simulations, the parameters, e, μ, p, i_1, ω_1, i_2, ω_2 and t, were better constrained to a smaller parameter space, the viewing angles were varied. First only one of the angles was changed, the other two remained fixed, later all were changed in combination. The variation of the viewing angles was performed systematically in steps of $10°$.

Changing the parameters and the combinations thereof resulted in a set of parameters describing the model that fits the observational data best (Fig. 3.6). This best fitting model was obtained by a visual inspection of the model fits and the observational data and by applying a set of different criteria (see also Fig. 3.9).

The set of Identikit model parameters delivering the best solution fitting of the HST image and the CO position-velocity diagram is comprised of the following parameters: The eccentricity at a value of $e = 1$. For the pericentric separation p a value of 0.125 delivered the best results. A mass ratio of the two galaxy disks of 1:2, an unequal merger (see Figs. 3.6 and 3.7) resulted in the best model fit. The lower mass galaxy (see Fig. 3.6 in dark grey) was initiated with an inclination i of $60°$ and a pericentric argument ω of $-15°$, the higher mass galaxy was described initially by an inclination i of $-60°$ and a pericentric argument ω of $40°$. The viewing angles of $\Theta_X = 90°$, $\Theta_Y = 180°$ and $\Theta_Z = 40°$ gave a good match of the simulation data to the observations. The evolution time scale for the merger resulted in a value of $t = 3$ (in internal units). 1σ errors of the Identikit modeling parameters are of the order of $8.3°$ for the viewing angles Θ_X, Θ_Y and Θ_Z. The errors for the internal time scale, the inclinations and pericentric arguments of the disks are $\Delta t \sim 0.1$, $\Delta i \sim 10°$ and $\Delta \omega \sim 10°$. A comparison of the obtained model fit of Arp 220 to the study of González-García & Balcells (2005), of merging

3.6. Identikit simulations

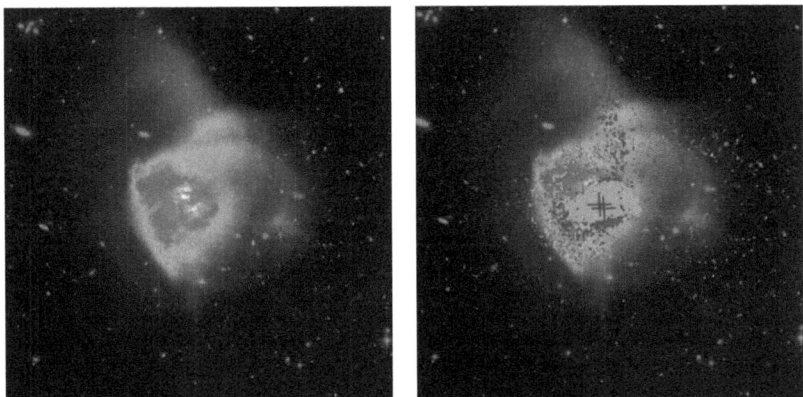

Figure 3.7.: Left: HST ACS/WFC F435W and F814W composite image of Arp 220 with an increased contrast in order to show the tidal tails of the galaxy more clearly. **Right:** The same, but superimposed with the x-y plane of the best model fit and also with an increased contrast (between model and background image).

galaxy systems, results in a translation of the evolution time t to a merger age of roughly $\sim 6 \times 10^8$ yrs. This value is in very good agreement with the age of the merging system determined by Mundell et al. (2001). They give a value of 7×10^8 yrs.

The comparison of the HST image and the best model fit in Fig. 3.7 shows that the model very well represents the properties of the Arp 220 in the HST observations. The sharp edge towards the north-east is extremely well represented by the model. The tidal tail of the less massive galaxy (in dark grey) is in almost perfect agreement with the extended emission towards the northwest. The more southern tidal tail is, at the basis towards the central region of Arp 220, in good agreement with the HST image. Further out this extended emission does not coincide with the model any more. As shown in Fig. 3.8, the overall shape of the model matches the CO p-v diagram. Note that a difference is that the CO p-v diagram was taken in the central region of Arp 220. Hence it does not cover the same velocity range as the simulations do and does not take the regions furthest outwards into account (see Fig. 3.8). The extensions to the NE and the SW side of the zero-offset axis of the p-v diagram describe the CO p-v diagram very well. The difference in the velocity range originates from the test particles further outward of the central region not covered by the CO data. Also the extended emission found

3. The Arp 220 merger on kpc scales

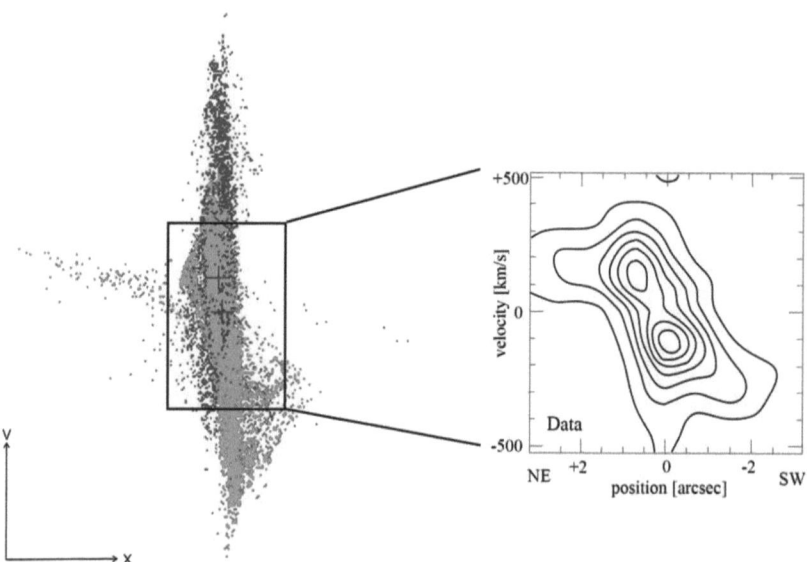

Figure 3.8.: Comparison of the position-velocity diagram obtained from the Identikit simulations (*left*) with the p-v diagram from CO observations (*right*). Note that the CO p-v diagram does not cover the same velocity range as the simulations do, since it was obtained for the center of Arp 220 and does not take the regions furthest outwards into account.

in the low resolution CO observations (Fig. 3.4) coincides very well with the simulations: At the location of the extended emission $\sim 10''$ to the south of the central part of Arp 220 a lot of emitting material is found in the Identikit simulations (Fig. 3.6).

The quality of the model fit was determined by the following criteria: Size and compactness of the merger galaxy, and by the location and position angle of the tidal tails. The position angles, α and β, of the tidal tails to the north-west (represented by the white line in Fig. 3.9) and closer to the central region (represented by the grey line crossing the white one in the same Figure) of Arp 220 have values of $\alpha \sim (124.6 \pm 9.6)°$ and $\beta \sim (6.9 \pm 7.6)°$. The sharp edge of the central region of Arp 220 is located at a position angle γ of $\sim (125.2 \pm 4.8)°$. The surface filling factor, a representation of the size and compactness of the galactic central bulge, was found to be $F \sim (79 \pm 6)\%$. This filling factor F represents the area within the box, shown in Fig. 3.9, occupied by the central region of Arp 220. F is determined by fraction of the area of the box and the box area, with the triangles subtracted where the

3.6. Identikit simulations

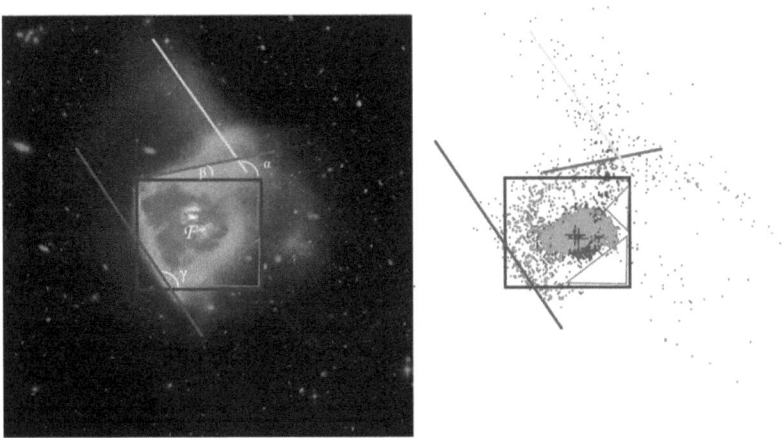

Figure 3.9.: The model evaluation criteria are shown on top of the HST image (*left*) and the best model fit (*right*) for a comparison of the observed data to the model fit and an assessment of the quality of this model. α, β and γ represent the position angles of the two tidal tails and the sharp edge of Arp 220 to the east, respectively. F is the surface filling factor, which is a measure for the size and compactness of the galaxy.

particle density is reduced to about zero:

$$F = \frac{box\ area - triangles}{box\ area}\ \% \qquad (3.2)$$

All errors stated for the results are errors at the 1σ level. By eye, this 1σ error for each criteria parameter was determined from the HST image. For the error values resulting thereof a χ^2 (Eq. 3.1) of 1.3 was derived by

$$\chi^2 = \sum_{i=1}^{N_{obs}} \frac{\left(f_{obs,i} - f_{model,i}\right)^2}{\sigma_i^2} \qquad (3.3)$$

as a measure for the deviation of the best model fit from the observed data. $f_{obs,i}$ is the respective criteria parameter measured in the observational data, $f_{model,i}$ is the parameter determined from the best model fit and σ_i is the observational error. Assuming $\Delta\chi^2 \sim 1$ for the variation of each Identikit parameter separately (one parameter was variable, the others were kept fixed), the internal errors for the Identikit model parameters were derived (see before). 10 model parameters in the Identikit tool are faced by 10 criteria parameters obtained from the observations: location (six parameters) and position angle (three parameters) of the tidal tails and the sharp edge of Arp 220, and the surface filling factor (one parameter). With the help of these criteria the quality of the fitted models was assessed. The comparison of the best model to the HST image in Fig. 3.9 shows how well it fits the observational data.

3.6.3. Discussion

Since it is known from observations at mm wavelengths that the mass ratio of the two nuclei Arp 220-West and Arp 220-East is about 1:1 (see e.g., Scoville et al., 1997; Sakamoto et al., 1999; Downes & Eckart, 2007), it is curious that the simulation results presented here give a value for the mass ratio μ of 1:2. One possible explanation is that the model, is not unique, i.e. the set of parameters fitting the HST and CO data. Nonetheless, the best model obtained here produces a 'fair' fit to the HST image and the CO data. Another point that speaks in the favor, of the model presented here, is the determined merger age that is in very good agreement with the literature value from Mundell et al. (2001).

Another point of critique may be the data that were used to fit the model to, since they were taken at two different wavelengths and are therefore basically tracing matter under different physical conditions. Only in the central arcseconds the gas has an impact on the overall gravitational potential, comparable to the stars impact on the potential. Further out the stellar potential clearly is dominating. Therefore the gas follows the potential of the stars and consequently the stars themselves. Since it was easier to guide the fits from the simulations by the tidal tails, seen in the HST image, more weight is put on the outer regions of the galaxy. Hence it is justifiable to use the HST data as well as the CO observations to fit the model results to. Outside the central region of Arp 220 gas is assumed to be found in clumps with sizes of about 1 pc. Therefore the gas clumps can be treated, to first order, as stars. Both, stars and gas clumps can then be handled, to first order, as collisionless particles. This fact is implemented in the Identikit code, so that it is truly justifiable to use this modeling tool for the purposes of the simulation presented.

3.6. Identikit simulations

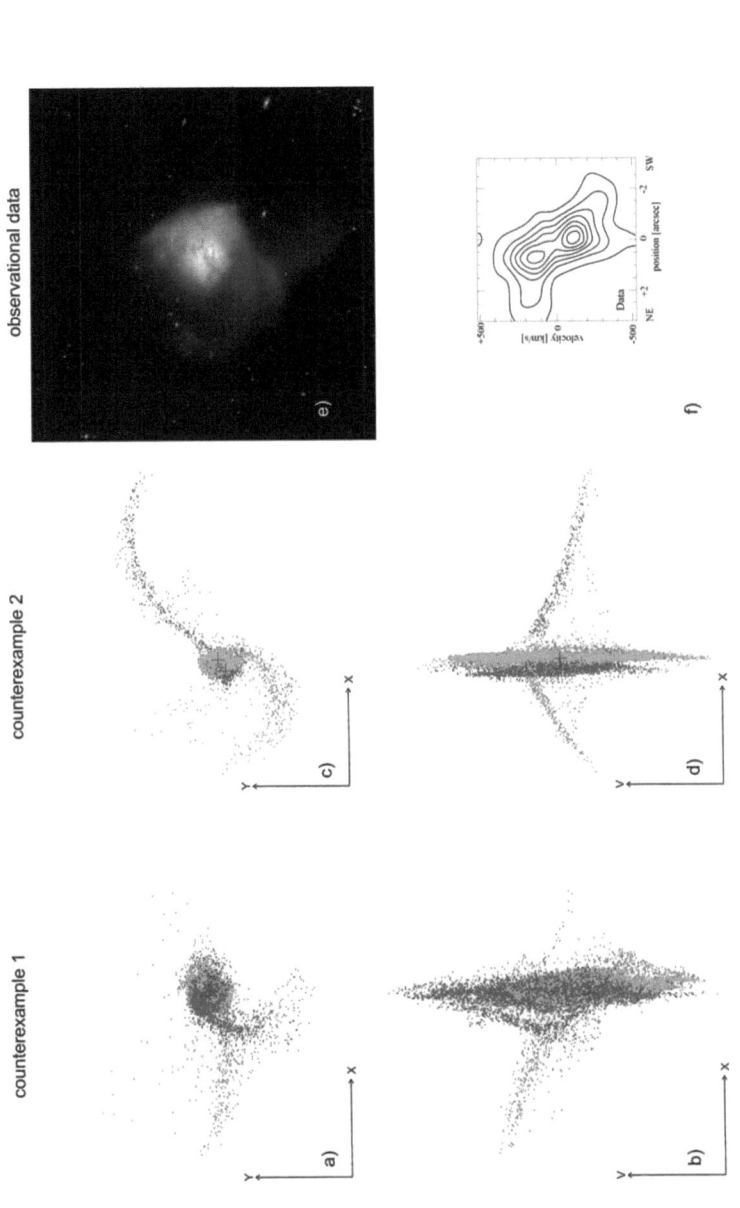

Figure 3.10.: Identikit simulation with the same parameter set as in the best model fit, but with a mass ratio μ of 1:1 (*left*) and a viewing angle Θ_X of 0° (*middle*). The x-y plane is shown at the top (a) and c)), the position-velocity diagram (x-v) at the bottom (b)and d)). The HST composite image and the CO position-velocity diagram (from Eckart & Downes, 2001) used to fit the Identikit simulations are shown in e) and f) for comparison. Note that the HST image is rotated by 180° to match the Identikit internal axis orientation.

For that particular reason the Identikit model can be used without restriction. To show that the model presented here is indeed the best one that could be found, two possible counterexamples are shown in Fig. 3.10. Since a mass ratio of 1:1 between the nuclei is commonly assumed, one counterexample representing this mass ratio is shown in Fig. 3.10 (*left*). It was modeled with the same set of parameters as for the best model fit, but with the different mass ratio μ. In this figure it is very clear that the model does not fit either the HST image nor the CO p-v diagram (Fig. 3.10). The same applies to the second counterexample (Fig. 3.10, *in the middle*). Here the same set of parameters as for the best model fit was used again, only this time with a viewing angle Θ_X of 0°, which corresponds to an angular deviation from the best fitting model of 10σ. The resulting model very clearly does not fit the underlying observational data.

With the help of the mass of the gas clumps and the total gas mass of the extended region to the south of Arp 220 the particle collision rate can be estimated. Using the critical density that is commonly found in gas clumps ($n_{crit} = 10^3 \ldots 10^5$ cm^{-3}), the clump mass (M_{clump}) can be derived by

$$M_{clump} = n_{crit} V m_H \; [M_\odot]. \qquad (3.4)$$

V is the volume of one clump (~ 1 pc^3), m_H (1.67×10^{-27} kg) is the mass of one H atom and M_{clump} is given in units of solar masses $M_\odot$. Since the gas mass of the southern region of the extended CO emission was calculated ($M_{gas} \sim 10^8 \; M_\odot$), the clump density can now be determined, via

$$M_{clump} \int N dV = M_{gas} \qquad (3.5)$$

where M_{clump} and M_{gas} are known, N is the clump density and dV is the volume of the southern region of extended CO emission. With N, the velocity dispersion σ (~ 10 km s^{-1}) and the cross-section s (~ 1 pc^2) known, finally the collision rate R_{coll} can be derived by:

$$R_{coll} = N \sigma s. \qquad (3.6)$$

From this value the time t_{coll} between two collisions is then determined by the reciprocal of the collision rate. This value gives an upper limit for t_{coll} of particles. Values for the time scale range between 3.6×10^8 and 3.6×10^{10} yrs, dependent on the critical density of the gas clumps used in the calculations. These time scales are comparable to the dynamic time scale of the merger itself. The other values for the time between collisions are even larger. That means the gas clumps in the outer regions can be treated as collisionless particles. Even if agglomeration between the gas clumps should occur, the energy contribution from the merger process, via tidal forces, to the velocity dispersion of the gas is large enough that the clumps in the outer regions do not collide any more. This supports that the Identikit modeling tool, that uses collisionless test particles in the simulations, is very well suited for my purposes.

Also the contribution of the atomic gas component can be neglected: Since Arp 220 is already in the final stage of merger the interacting galaxy component particles already collided, in the center of Arp 220, at least 1–2 times with each other. Therefore the HI is either stripped from the region studied with the merger simulations (Fig. 3 of Hibbard et al., 2000) or is already transformed into molecular gas (H_2).

3.7. Conclusions

In this chapter on Arp 220 the results on the study of the interferometric CO data and the Identikit simulations of the merger are discussed. The eastern nucleus of Arp 220 is studied and the object as a whole is studied for more extended structure in CO gas emission. The analysis of the CO lower resolution data shows that the CO-SE peak is a very promising candidate to be the 'real', but deeply embedded/highly dust obscured, nucleus of Arp 220-East. In addition to that Arp 220-East shows a rotational motion that corresponds to the overall rotation of the underlying molecular disk. Indications for emission $\sim 10''$ towards the south, as well as to the north and to the west of the two nuclei were found in the low resolution $CO\,(1-0)$ maps. Furthermore simulations of the merger in Arp 220 were performed with the Identikit modeling tool. The model parameters describing the galaxy merger best give a mass ratio of 1:2 and result in a merger of $\sim 6\times 10^8$ yrs, which is in good agreement with values from the literature, such as Mundell et al. (2001). Apparently Arp 220 can be reckoned a very interesting case of merger galaxy, where the object can be placed between a minor merger (from the CO observations) in one extreme and an equal (1:1) merger (from the simulations, CO data and published models in the literature) in the other.

4. HI in nearby low-luminosity QSO host galaxies

4.1. Introduction

Molecular gas is the fuel for any activity process, whether it is in the form of vigorous bursts of star formation or an intensively accreting super massive central black hole. The study of its morphology and kinematics is of key importance to understand the fueling mechanisms that keep the activity alive over cosmologically significant time scales. As important as studying the ongoing activity processes is studying the trigger that leads to it in the first place. It is assumed that galaxy interaction thereby plays an important role, initializing the transport of gas from the outer parts of the involved galaxies into their centers and so igniting the starburst and/or AGN activity. Sanders et al. (1988) already suggested early on that there might be an evolutionary sequence between the onset of starburst and AGN activity with starbursts as precursors for AGN. This hypothesis is mainly based on the remarkable resemblance between luminous starburst galaxies (so called Ultra-Luminous-Infrared-Galaxies, or ULIRGs) and galaxies with a pronounced AGN signature (so called QSO host galaxies), such as their enormous amounts of molecular gas. However, while ULIRGs often exhibit significant signs of galaxy interaction, this is not necessarily true for QSO host galaxies. In fact, optical studies of QSO host galaxies (e.g., Bahcall et al., 1997; Boyce et al., 1998) find the majority to be rather morphologically normal without any clear signs of a past or ongoing interaction, though $\sim$50% of their sample live with (projected) nearby companions. Much effort has been invested in studying the molecular gas emission in QSO host galaxies (e.g., Scoville et al., 2003;

Krips et al., 2006a,b; Bertram et al., 2007). Atomic hydrogen, on the other hand, tracing the cold atomic gas, may be a much better indicator for galaxy interaction. Only little is known so far on the atomic gas content in QSO host galaxies. Lim & Ho (1999), Lim et al. (2001) and Ho et al. (2008a,b) present one of the first studies of HI emission in nearby QSO host galaxies. They find that while most of their observed sources show signs of a past or ongoing tidal interaction, they do not yet physically merge with another galaxy. This is in contrast to the proposed sequence by Sanders et al. (1988) in which QSOs are thought be the merging or merged descendants of ULIRGs. In this chapter the study of neutral atomic hydrogen in 27 nearby low-luminosity QSOs is presented. The observations are part of an ongoing project aimed at enhancing the statistics on nearby QSOs.

In Sect. 4.2 the sample selection is described, Sect. 4.3 reports on the observations and data reduction. The description of the analysis, the results of the Effelsberg and the VLA observations, and the conclusions/discussion follow in Sect. 4.4, 4.5, 4.6 and 4.7.

Unless otherwise stated, $H_0 = 75$ km s^{-1} Mpc^{-1} and $q_0 = 0.5$ are assumed throughout the thesis.

4.2. The sample

The sample of type 1 QSOs described in this chapter of the thesis is volume limited. As the sole selection criterion an upper redshift limit of $z < 0.06$ was applied. This value has been set to ensure the observability of the important diagnostic CO(2−0) rotation vibrational band head absorption line in the K-band. The sources have been selected from the Hamburg/ESO survey of optically bright QSOs (HES; Wisotzki et al., 2000). These 27 sources, with a declination $\delta > -30°$ (by means of the observability from the Effelsberg 100-m telescope), have been searched for CO emission with the IRAM 30 m telescope on Pico Veleta (Spain) and SEST (Swedish ESO-Submillimeter Telescope) on La Silla (Chile). 27 objects (Table 4.1) were

4.2. The sample

Table 4.1.: List of sources observed at 21 cm

Object	RA(2000) [h][m][s]	DEC(2000) [°][′][″]	Seyfert[a] Type	Morphological[b] Type	v_0 (LSR) [km s^{-1}]	z	D_L [Mpc]	$t_{obs,on}$[c] [min]	Cali-[d] bratores	Resolution[e] [km s^{-1}]	rms noise [Jy]
HE 0021−1819	00:23:55.3	−18:02:50		E, C	15954	0.053	215.4	60	1, 1	1.3	0.008
HE 0040−1105	00:42:36.8	−10:49:21	Sy 1.5	E, C	12578	0.042	169.6	60	1, 1	0.3	0.013
HE 0045−2145	00:47:41.3	−21:29:27		S	6403	0.021	86.0	60	2, 1	5.2	0.004
HE 0114−0015	01:17:03.6	+00:00:27		E, C, poss. M	13682	0.046	184.3	90	2, 1, 1	1.3	0.006
HE 0119−0118	01:21:59.8	−01:02:25	Sy 1.5	E	16412	0.055	221.5	90	1, 1	2.1	0.007
HE 0150−0344	01:53:01.4	−03:29:24		M	14329	0.048	193.3	90	1, 3, 1	16.5	0.004
HE 0212−0059	02:14:33.6	−00:46:00	Sy 1.2	E, C	7921	0.026	106.2	60	2, 3	8.2	0.004
HE 0224−2834	02:26:25.7	−28:20:59		M, C	18150	0.060	245.4	60	1, 1	8.0	0.003
HE 0227−0913	02:30:05.4	−08:59:53	Sy 1	S	4914	0.016	65.8	60	2	2.6	0.007
HE 0232−0900	02:34:37.7	−08:47:16	Sy 1	R, M, C	12886	0.043	173.7	30	2, 1	32.0	0.002
HE 0253−1641	02:56:02.6	−16:29:16	Sy 2	S	9580	0.032	128.9	30	1	0.3	0.019
HE 0433−1028	04:36:22.2	−10:22:33	Sy 1	S	10651	0.036	143.1	60	1, 1	2.1	0.007
HE 0853−0126	08:56:17.8	−01:38:07	Sy 1	E, C	17899	0.060	242.1	60	1, 1	0.3	0.017
HE 0949−0122	09:52:18.9	−01:36:44	Sy 1.5	E	5905	0.020	79.1	60	4, 3	4.1	0.016
HE 1011−0403	10:14:20.6	−04:18:41	Sy 1	S, C	17572	0.059	237.6	30	1	0.3	0.021
HE 1017−0305	10:19:32.9	−03:20:15	Sy 1	S, C, poss. M	14737	0.049	199.0	30	1	0.3	0.018
HE 1029−1831	10:31:57.3	−18:46:34	Sy 1	S, C	12112	0.040	163.1	60	3, 1	16.0	0.001
HE 1107−0813	11:09:48.5	−08:30:15		E	17481	0.058	236.3	30	1	0.3	0.021
HE 1108−2813	11:10:48.0	−28:30:03	Sy 1.5	S	7198	0.024	96.5	90	4, 1, 1	0.3	0.017
HE 1126−0407	11:29:16.6	−04:24:08	Sy 1	S, C	18006	0.060	243.7	30	1	0.3	0.020
HE 1237−0504	12:39:39.4	−05:20:40	Sy 1	S	2531	0.008	33.6	30	4	4.1	0.008
HE 1248−1356	12:51:32.4	−14:13:17	Sy 1	S, C	4338	0.015	58.2	60	4, 3	2.1	0.002
HE 1330−1013	13:32:39.1	−10:28:53	Sy 1	S, C	6744	0.023	90.4	90	1, 3, 3	16.5	0.006
HE 1353−1917	13:56:36.7	−19:31:14	Sy 1	S	10472	0.035	140.7	60	1, 1	0.3	0.017
HE 2222−0026	22:24:35.3	−00:11:04		E	17414	0.058	235.5	30	1	0.3	0.019
HE 2233−0124	22:35:41.9	+01:39:33	Sy 1	S, C	16913	0.056	228.5	30	1	0.3	0.018
HE 2302−0857	23:04:43.4	−08:41:09	Sy 1.5	S, C	14120	0.047	190.4	90	1, 1, 1	8.0	0.002

[a] Taken from the NED.

[b] E denotes elliptical morphology; S represents a spiral morphology; M stands for possible mergers or merger remnants; R denotes ringed objects; C marks galaxies with other extragalactic sources within a projected distance of up to 340 kpc.

[c] Total integration time spent on the source.

[d] 1 – Holmberg I; 2 – UGC 2345; 3 – Sextans B; 4 – NGC 4303.

[e] Spectral resolution of the smoothed spectra shown in Fig. 4.1.

detected in ^{12}CO(1−0) and ^{12}CO(2−1) (see Bertram et al., 2007). They feature recession velocities are in the range 2 500 km s^{-1} < v < 18 000 km s^{-1}. Bertram et al. (2007) confirmed that the majority of galaxies hosting low-luminosity QSOs are rich in molecular gas. In 2007 and 2008 the HI sub-sample was searched for 21 cm HI emission. This sample complements the study of Ho et al. (2008a), but contrary to their sample, the sources from the HI sample have positions further to the south.

A more detailed description of the 'nearby QSO sample' can be found in Bertram et al. (2007).

4.3. Observations and data reduction

The observations of the HI 21 cm (v = 1420.40575 MHz) emission line were carried out with the Effelsberg 100-m telescope in 2007 and 2008. The 18–21 cm (1.29–1.72 GHz) two-channel L-band HEMT facility receiver was used as the frontend, in conjunction with the 8192-Channel-Autocorrelator (AK 90) and the Fast Fourier Transform Spectrometer (FFTS) as backends. With bandwidths of 10 MHz (AK 90) and 20 MHz (AK 90 + FFTS), centered on the redshifted rest frame frequency for each galaxy, the observational data show velocity ranges Δv from $\sim$ 4 200 to $\sim$ 21 000 km s^{-1} and a velocity resolution between 0.2577 and 2.061 km s^{-1}, depending on the backend and settings. The measurements were done in position-switching mode with 5 minutes on and 5 minutes off the source positions, with an off-position 45′ away. This provides the advantage of better baselines and the reduction of atmospherical influences. The beam efficiency was $\sim$ 1.15 for a beam size of 9.5′ at 21 cm. The on-source integration times range between 30 and 90 minutes. Daily pointing checks were performed using sources from the Effelsberg Catalog of pointing and flux density calibration (Ott et al., 1994). As flux calibrators Holmberg I, NGC 4303, Sextans B and UGC 2345 (fluxes taken from Tifft & Huchtmeier 1990 and from Springob et al. 2005) were used.

4.3. Observations and data reduction

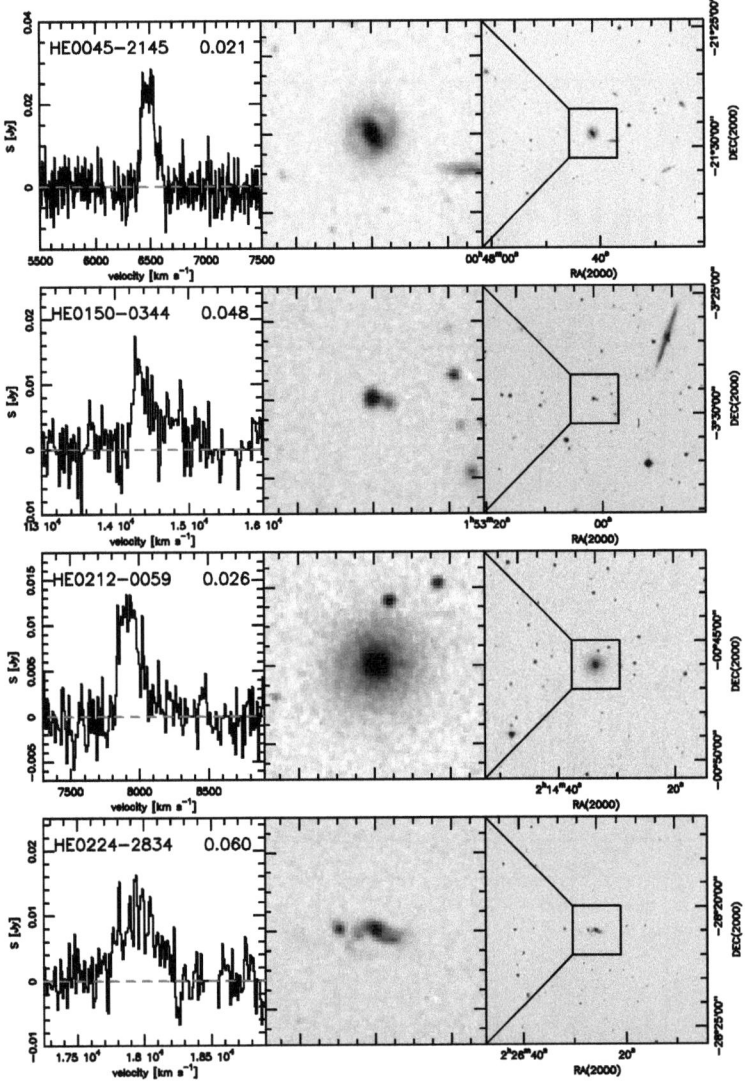

Figure 4.1.: HI spectra of detected host galaxies from the 'nearby QSO sample' observed with the Effelsberg 100-m telescope and optical DSS images of objects detected in HI. The images in the middle extend over $2'$ and the ones to the right contain $9.5'$, which is roughly the size of the beam at 21 cm. North is up and East to the left. Each source is identified by its HES name (top left corner of the spectrum) and the redshift (top right corner). The spectral resolution of the spectra in this figure range from 0.3 km s^{-1} up to 32 km s^{-1}. If regions in the spectrum are effected by RFI (Radio Frequency Interference) and are not used for baseline-fitting and noise estimates, they are set to zero intensity.

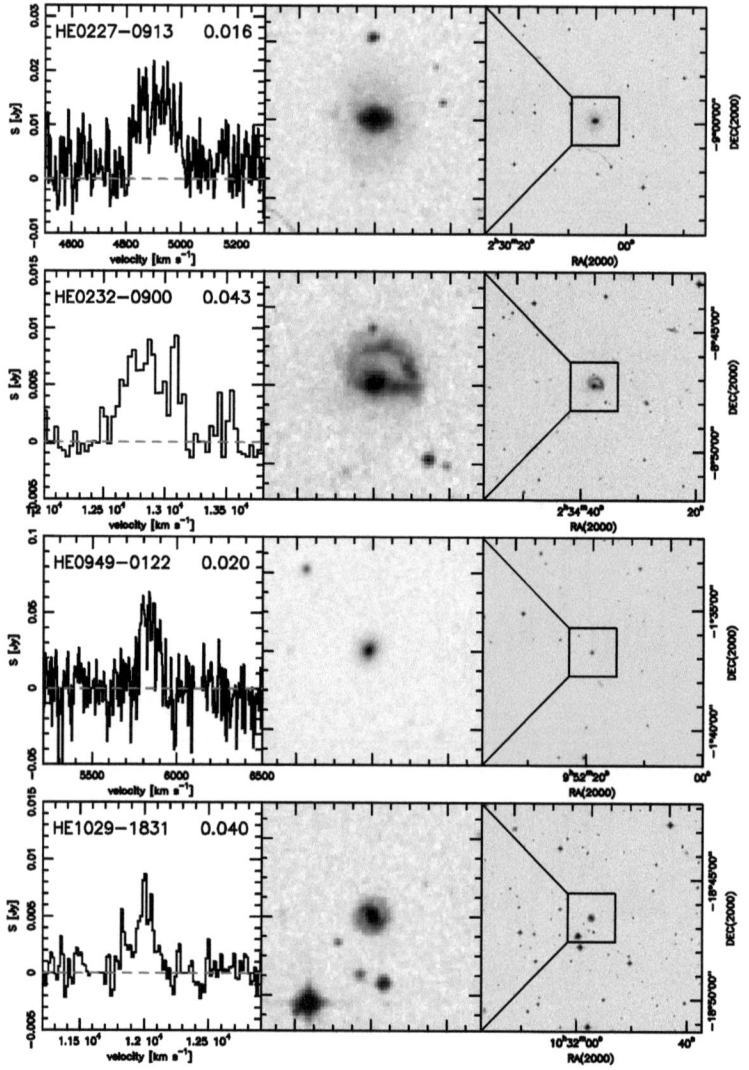

Figure 4.1.: continued

4.3. Observations and data reduction

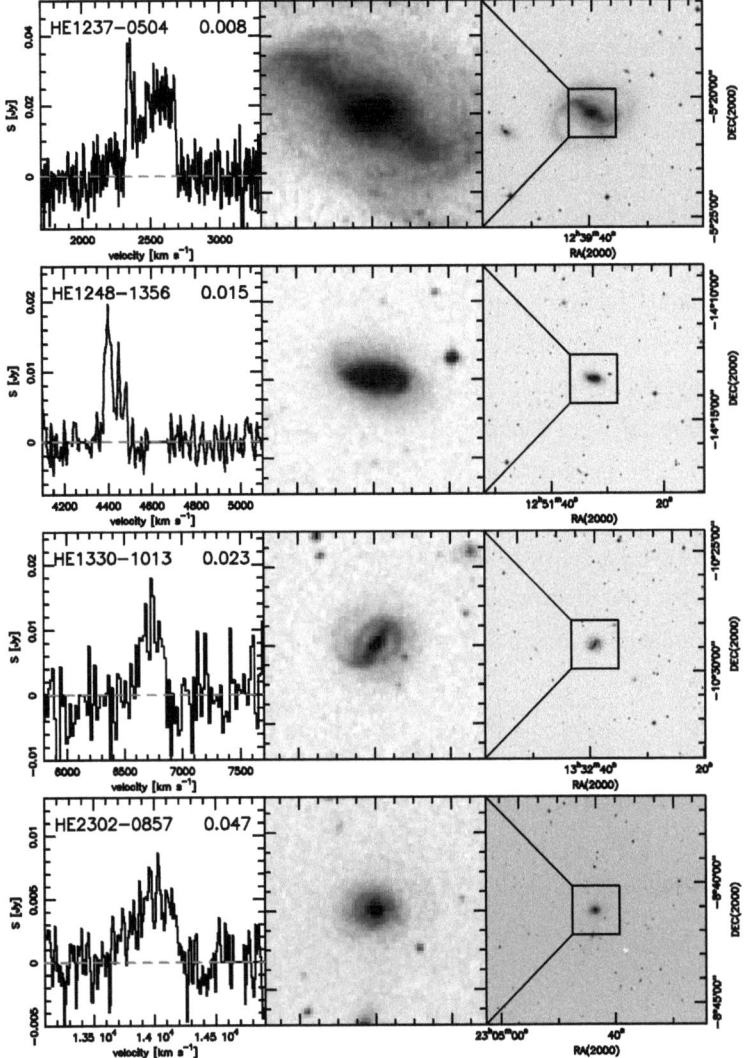

Figure 4.1.: continued

The Effelsberg 100-m telescope site is known for exhibiting low level radio frequency interference (RFI) in the 21 cm band. However as the RFI signals are situated well outside the range of the expected central velocity or can be filtered out, these disturbances have no influence on the observational results. The data were reduced and analyzed with the CLASS package of IRAM's GILDAS software. All spectra have been averaged, when applicable. To obtain a better signal-to-noise ratio, most of the spectra were Hanning smoothed. For data taken on different dates the subscans of each observation were calibrated with respect to a suitable flux calibrator (see Fig. 4.1 and Table 4.1). Baselines were fitted and subtracted from each subscan individually before resampling and averaging of the spectra. Another baseline fit was performed after the final resampling and averaging of all available data for the source. All polynomial fits to the baselines have been of the order of one. The intensity errors ΔI were determined following the procedure from Bertram et al. (2007). The geometric average of the line error $\Delta I_L = \sigma \, v_{res} \sqrt{N_L}$ and the baseline error $\Delta I_B = \sigma \, v_{res} \, N_L / \sqrt{N_B}$ were taken into account. σ is the rms noise in Jy, v_{res} the spectral resolution in km s^{-1}, N_L the number of channels over which the line spreads and N_B the number of channels used for fitting a polynomial to the baseline.

4.4. Analysis

The galaxies observed during this survey do not form a complete sample, but the results can still be used to improve the knowledge about the population of nearby AGNs. In order to derive the atomic neutral hydrogen gas mass M_{HI}, the following formula (e.g., Shostak, 1978) was applied:

$$M_{HI} = 2.36 \times 10^5 D_L^2 \int S_v \, dv \, M_\odot. \tag{4.1}$$

D_L denotes the luminosity distance in Mpc and $\int S_\nu d\nu$ the flux integral in Jy km s^{-1}. For a comparison the dynamical mass was determined via

$$M_{\mathrm{dyn}} = \frac{v_{\mathrm{HI}}^2 r}{G} M_\odot \qquad (4.2)$$

(v_{HI} is the velocity width from which 90% of the HI line emission arises and r is the estimated HI radius) under the assumption of a disk geometry in virial equilibrium, which is appropriate for gas rich objects. Since the size of the HI extension is not known the area of the sources was estimated from the DSS images (Fig. 4.1) by circular apertures. These apertures were centered on the nucleus of each galaxy. The optical radius was taken as the value where the counts of the galaxy emission in the DSS image were 3σ larger than the background. This radius was multiplied by the factor for the median $r_{\mathrm{HI}}/r_{\mathrm{opt}}$ of 1.50 from Haan et al. (2008), and assuming a circular area. The area was then derived by $A = \pi r^2$ which in turn was used for the determination of the dynamical mass. As expected the values for M_{dyn} are up to two orders of magnitude larger than the HI masses determined from the integrated intensity (Table 4.2). The infrared luminosity was determined from the IRAS fluxes (12, 25, 60 and 100 μm fluxes were taken from the IRAS Faint Source Catalog, Moshir et al., 1990) and with the following formulae (Sanders & Mirabel, 1996):

$$F_{\mathrm{IR}} = 1.8 \times 10^{-14} \left(13.48 f_{12} + 5.16 f_{25} + 2.58 f_{60} + f_{100} \right) \frac{W}{m^2} \qquad (4.3)$$

$$L_{\mathrm{IR}} = 4\pi D_L^2 F_{\mathrm{IR}} \, L_\odot. \qquad (4.4)$$

These results are summarized in Table 4.2. Given are the integrated line intensity I_{HI} (column 3), the line width at half maximum intensity W_{50} (col. 4), an estimate on the line shape (col. 5), the atomic neutral hydrogen mass M_{HI} (col. 6), the dynamical mass deduced from the HI data (col. 7), the total hydrogen gas mass of the host galaxy (col. 8), as well as the molecular to atomic gas mass fraction (col. 9) and the infrared luminosity L_{IR} (col. 10).

4. HI in nearby low-luminosity QSO host galaxies

Table 4.2.: Summary of the HI properties

Object	v_0 (LSR) [km s^{-1}]	$I_{\rm HI}$[a] [Jy km s^{-1}]	W_{50}[b] [km s^{-1}]	LS[c]	$M_{\rm HI}$ [$10^9 M_\odot$]	$M_{\rm dyn}$ [$10^{10} M_\odot$]	$M_{\rm tot,gas}$ [$10^{10} M_\odot$]	$L'_{\rm CO}/M_{\rm HI}$ [$\frac{\rm K\,km\,s^{-1}\,pc^2}{M_\odot}$]	$L_{\rm IR}$ [$10^{10} L_\odot$]	f	Σ_M [$M_\odot$ pc^{-2}]	$M_{\rm HI}/L_{\rm IR}$ [$M_\odot/L_\odot$]	$M_{\rm HI}/L_B$ [$M_\odot/L_\odot$]
HE 0045	6 403	3.29±0.13	180.8	⊓A	5.7±0.2	7.1	11.5	0.13	8.10	0.6	20.0	0.08	1.2
HE 0150	14 329	2.70±0.28	263.0	⊓A	23.8±2.5	45.0	26.8	0.02	11.10	0.9	27.0	0.21	4.9
HE 0212	7 921	1.97±0.18	206.4	⊓⊓	5.3±0.5	3.3	11.7	0.09	4.07	1.1	121.6	0.13	0.4
HE 0224	18 150	2.64±0.26	271.0	⊓A	37.5±3.7	90.8	41.5	0.01	10.50	0.3	20.3	0.34	3.8
HE 0227	4 914	2.58±0.16	193.9	⊓	2.6±0.2	0.9	3.4	0.07	1.08	1.0	466.9	0.24	0.7
HE 0232	12 886	3.31±0.40	559.0	∧∧	23.6±2.8	46.9	56.6	0.18	18.50	0.8	114.9	0.13	0.8
HE 0949	5 974	5.88±0.46	181.4	∧∧	8.7±0.7	3.5	9.5	0.01	58.0	1.5	58.0	0.15	1.7
HE 1029	12 112	1.05±0.10	241.0	∧	6.6±0.6	17.2	17.2	0.20	21.20	0.9	14.4	0.05	0.6
HE 1237	2 531	7.55±0.36	355.5	⊓⊓	2.0±0.1	67.5	2.82	0.05	1.44	0.9	1.00	0.14	1.2
HE 1248	4 338	0.78±0.03	119.5	∧∧	1.2±0.2	5.7	2.2	0.08	1.29	1.4	2.6	0.12	3.5
HE 1330	6 744	1.84±0.35	180.4	⊓⊓	3.6±0.7	16.1	5.0	0.05		1.0	6.3		1.6
HE 2302	14 120	1.39±0.18	338.0	⊓A	11.9±1.5	130.7	19.7	0.08		1.3	4.7		1.1
Non-detections													
HE 0021	15 889	< 2.03	222.7		< 22.2		< 2.4	< 0.01					< 4.1
HE 0040	12 390	< 1.58	222.7		< 10.7		< 1.3	< 0.03					< 4.2
HE 0114	13 672	< 1.75	222.7		< 14.0		< 1.7	< 0.03					< 2.5
HE 0119	16 291	< 1.42	222.7		< 16.5		< 2.7	< 0.08					< 1.0
HE 0253	9 470	< 2.58	222.7		< 10.1		< 1.5	< 0.06	24.6			< 0.07	< 1.0
HE 0433	10 658	< 2.21	222.7		< 10.7		< 2.9	< 0.21	4.90			< 0.21	< 2.2
HE 0853	17 930	< 1.03	222.7		< 14.3		< 2.1	< 0.06	22.7			< 0.05	< 1.1
HE 1011	17 388	< 0.56	222.7		< 7.5		< 1.5	< 0.12					< 1.5
HE 1017	14 985	< 1.22	222.7		< 11.4		< 1.6	< 0.05				< 0.12	< 0.2
HE 1107	17 088	< 1.03	222.7		< 13.6		< 1.8	< 0.04	9.88				< 1.2
HE 1108	7 199	< 0.50	222.7		< 1.1		< 0.9	< 0.85					< 0.8
HE 1126	17 988	< 1.83	222.7		< 25.7		< 3.4	< 0.04	10.4			< 0.01	< 0.1
HE 1353	10 388	< 2.44	222.7		< 11.4		< 1.7	< 0.07					< 0.5
HE 2222	17 431	< 1.22	222.7		< 15.9		< 1.7	< 0.01					< 4.7
HE 2233	16 933	< 2.93	222.7		< 36.1		< 4.0	< 0.01					< 3.7
													< 4.4

[a] Upper limits to the flux. Errors represent 3 σ values.
[b] For the non-detections, values of W_{50} correspond to the median value of all detected sources.
[c] Line shape: ∧ represents a triangular shaped line profile; A stands for asymmetries in the line profiles; spectra with ⊓ show a boxy line profile; ⊓⊓ mark line profiles of a double-horn shape.

4.5. Results and discussion

In Fig. 4.1 the obtained HI spectra are shown together with the associated DSS images of the region with a diameter of $2'$ and $9.5'$ around the source position. The images should serve as a means to estimate the source confusion. The spectrum of HE 1248–1356 is remarkable in the sense that the velocity range 4 200–4 350 km/s shows strong CO emission (see Bertram et al., 2007) but no HI emission. As this source has a close companion (see the DSS images), it is possible that the HI gas was partially more disturbed/stripped/disrupted, by the companion. This could cause the 'one-sidedness' of the spectrum indicating that the part of the host galaxy moving towards us has less neutral atomic hydrogen than the one receding. Why wasn't the CO affected? It could just be that the disturbance was only affecting the more outwards lying HI gas and not yet the molecular CO in former times. Since then HI was converted into H_2 (CO) and the HI line in its present shape could be a remnant signature. The coordinate labels of RA and DEC (J2000), as well as the flux in Jy and the velocity in km s^{-1}, are indicating the scales. Each source name is denoted above the spectrum. The HI properties and results are summarized in Table 4.2: Integrated line intensities cover a range from 0.5 up to 7.6 Jy km s^{-1}. The intensity upper limits and errors for non-detections represent 3σ values. The line widths (W_{50}) are of the order of hundreds of km s^{-1}. The sources HE 0232–0900 ($W_{50} = 559$ km s^{-1}) and HE 2302–0857 ($W_{50} = 338$ km s^{-1}) are two special cases: They both have extraordinary large line widths in CO (FWZI: 597 km s^{-1} and 653 km s^{-1}, from Bertram et al., 2007) and HI emission. The HI line emission detected from HE 2302–0857 is even more extraordinary when comparing the full width at zero intensity (FWZI) of the HI line, with a width of 616 km s^{-1}, which is comparable to the line width (FWZI) in CO ($\Delta v_{CO} = 653$ km s^{-1}), although the galaxy is not seen edge-on. For HE 0232–0900, however, the HI line is narrower than the CO line. This indicates that the outer regions of the galactic disk were stripped from HI, or are simply depleted from the atomic gas component. As the DSS images (Fig. 4.1) show a peculiar morphology of

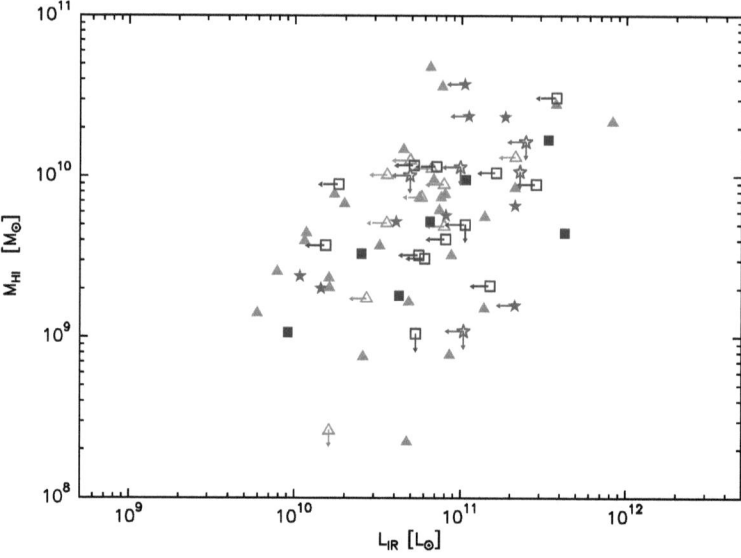

Figure 4.2.: HI mass as a function of infrared luminosity for the nearby QSOs. Stars represent the sample from this work. Squares and triangles represent sources taken from Ho et al. (2008a) for getting better statistics: squares represent the sources from the Ho sample itself and triangles represent a literature sample assembled by Ho et al. (2008a). Filled squares and triangles mark available values for L_{IR}, whereas open ones with arrows represent upper limits of L_{IR} or M_{HI}. Filled stars, for the sample from this work, represent HI and CO detected sources and unfilled ones denote upper mass limits for the CO but not HI detected sources. Arrows show upper limits for mass and infrared luminosity.

this object it is also possible that the current geometry is due to a previous interaction/merger with another galaxy.

The total neutral hydrogen masses for the sample sources range between 1.1 and 37.5×10^9 $M_\odot$, whereas the dynamical masses, that represent upper HI mass limits, have values from 0.9×10^{10} up to 1.3×10^{12} $M_\odot$. Considering the upper limits for the non-detections the median HI mass value changes to 11.4×10^9 $M_\odot$. Taking this number for the whole HI subsample into account, the atomic hydrogen mass is a factor of two higher for this works sample sources than the HI mass of the Milky Way ($M_{HI} = 5.5 \times 10^9$ $M_\odot$, see Hartmann & Burton, 1997). Comparing the hydrogen mass for the QSOs in the HI subsample to ULIRGs (10^9 to few 10^{10} $M_\odot$; Mirabel & Sanders, 1988; Lonsdale et al., 2006) gives rise to a mass ratio of 1:1.

Fig. 4.2 shows a plot of the atomic hydrogen mass versus infrared luminosity containing sources from this work and the work of Ho et al. (2008a). In comparison to Ho et al. (2008b) the work in this thesis is focused more on the radio-IR properties, rather than the optical HI relations. The position of the objects in this graph are in good agreement with the work of Young et al. (1989) about infrared bright galaxies: In general, galaxies with high IR luminosities have high neutral atomic hydrogen masses. Although their work places an emphasis on even more nearby sources ($D_L = 36.3$ Mpc), the mean luminosity distance for the enlarged sample (this work + Ho et al., 2008a) plotted here was 219.5 Mpc. The comparison of these results shows no evolutionary effect in redshift due to M_{HI} or L_{IR} (see also Table 4.3 and Haan et al., 2008). The comparison between the molecular gas content (taken from Bertram et al., 2007) and the atomic one results in a median fraction of $L'_{CO}/M_{HI} = 0.06$ for this work's sample (including the upper limits for the non-detections). Maiolino et al. (1997) measured a value of 1.79 for their M_{H_2}/M_{HI} ratio (for galaxy types up to Sy 1.5). Since it was not sure whether Bertram et al. (2007) and Maiolino et al. (1997) used the same CO-to-H_2 conversion factor, I simply compare the L'_{CO}-to-M_{HI} ratios. The median L'_{CO}/M_{HI} value given by the data of Maiolino et al. (1997) results in 0.14, which is about a factor of two higher than the value obtained for the sample in this work. Since the median CO luminosity L'_{CO} is about the same in both samples, $5.55 \times 10^8 \, M_\odot$ for Maiolino et al. (1997) and $5.40 \times 10^8 \, M_\odot$ (CO data taken from Bertram et al., 2007), the galaxies in their sample are less rich in atomic hydrogen than the ones in this study. In a more recent publication, Haan et al. (2008) compiled a study of HI properties in AGN host galaxies taken from the NUGA sample. These sources, which lie a step lower in redshift range ($D_L = 6.6 \ldots 53.9$ Mpc, see Table 4.3), are compared with the HI subsample. They find more ring structures and dynamically disturbed HI disks in LINERs than in Seyfert host galaxies. Since the HI subsample consists only of Seyfert galaxies, the results from this work just account for part of the Haan sample.

The distribution of data points in Fig. 4.2 shows that the HI data points sample an upper envelope of the M_{HI}/L_{IR} distribution. The sources detected

Table 4.3.: Sources from Haan et al. (2008)

Object[a]	RA(2000)[a] [h] [m] [s]	DEC(2000)[a] [°] ['] ['']	v_{hel}[a] [km s^{-1}]	D_L[a] [Mpc]	HI line width[a] [km s^{-1}]	CO line width [km s^{-1}]	CO-Ref.[b]	f	L_{IR} [$10^{10} L_\odot$]	Σ_M [$M_\odot$ pc^{-2}]
NGC 3147	10:16:53.65	+73:24:02.7	2820	40.9	430	420	1	0.98	21.5	1.81
NGC 3627	11:20:15.03	+12:59:29.6	727	6.6	431	409	2	0.95	86.6	2.82
NGC 4569	12:36:49.80	+13:09:46.3	-235	16.8	396	385	3	0.97	21.2	1.34
NGC 4826	12:56:43.69	+21:40:57.5	408	4.1	372	373.75	4,5,6,7	1.00	86.8	0.43
NGC 5953	15:34:32.39	+15:11:37.7	1965	36	349	230	1	0.66	25.0	0.94
NGC 6574	18:11:51.23	+14:58:54.4	2282	38	340	350	1	1.03	35.7	1.88
NGC 6951	20:37:14.09	+66:06:20.3	1424	24.1	338	380.2	8	1.12	34.8	1.71
NGC 1961	05:42:04.80	+69:22:43.3	3934	53.9	767	731	9	0.95	20.1	12.84
NGC 3368	10:46:45.74	+11:49:11.8	897	8.1	399	365	10	0.91	24.4	1.33
NGC 3718	11:32:34.85	+53:04:04.5	994	17.0	465	470	11	1.01	5.38	1.74
NGC 4321	12:22:54.90	+15:49:20.6	1571	16.8	291	200	12	0.69	50.5	2.91
NGC 4579	12:37:43.52	+11:49:05.5	1519	37	374	380	13	1.02	15.1	1.26
NGC 4736	12:50:53.06	+41:07:13.7	308	4.3	284	270	5	0.95	126	1.53
NGC 7217	22:07:52.38	+31:21:33.4	952	16.0	326	320	14	0.98	15.8	2.13
NGC 7782	09:14:05.11	+40:06:49.2	2562	37.3	246	260	1	1.06	20.8	3.29
NGC 5248	13:37:32.07	+08:53:06.2	1153	15	311	300	5	0.96	45.7	2.33

[a] taken from Haan et al. (2008)

[b] References: 1 - Young et al. (1995); 2 - Sage (1993); 3 - Boselli et al. (1995); 4 - Helfer et al. (2003); 5 - Nishiyama & Nakai (2001); 6 - Casoli & Gerin (1993); 7 - Aalto et al. (1991); 8 - Kohno et al. (1999); 9 - Tutui & Sofue (1997); 10 - Braine et al. (1993); 11 - Pott et al. (2004); 12 - Sakamoto et al. (1995); 13 - Sofue et al. (2003); 14 - Combes et al. (2004)

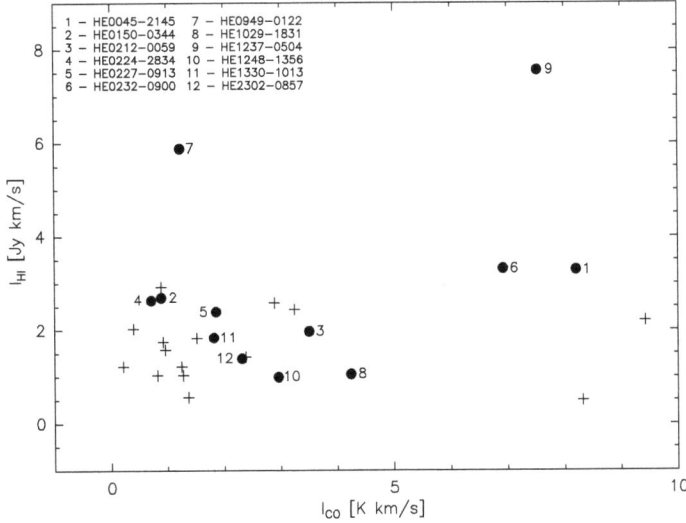

Figure 4.3.: HI flux vs. CO intensity (taken from Bertram et al., 2007). Crosses mark the upper limits of HI non-detected sources. Filled circles represent the intensities for detected sources. Each object is identified by a number, which is related to the source name given in the figure legend.

in HI emission are those that have the largest amount of atomic hydrogen, compared to the molecular gas mass, in their hosts. This could be an explanation for the lack of more HI detections in this work. Upper limits of the non-detected sources show, that with deeper integrations, one would be able to correct for this deficiency in the observations.

In Fig. 4.3 the HI and the CO intensities of the detected sources in HI subsample are compared. Similar to Fig. 4.2 it is obvious from Fig. 4.3 that the data points at hand sample an upper envelope of the HI mass distribution. If only the certain HI detected sources from this work are considered then a high HI mass is present almost independently of the molecular mass content. Remarkable about this plot is that the two sources with the largest CO intensity values (HE 0433–1028 and HE 1108–2813) are not detected in atomic hydrogen. This could mean that most of the available neutral atomic gas was already converted to molecular gas, i.e. H_2 (CO). The number of upper limits in the HI line strength is larger for galaxies with low molecular gas content. This implies that a correlation may be present, such that low molecular gas mass also implies a low atomic gas mass. On a statistical basis, the HI detection fraction above and below the median value of the molecular gas masses (taken from Bertram et al., 2007) could give a hint for the confidence limit of this trend in this chapters sample. The median value results in $M_{H_2} = 2.10 \times 10^9\ M_\odot$. Then the distribution of the HI masses of the sources above and below the median of the molecular gas mass was studied. The median HI mass above the median molecular gas mass is $M_{HI} = (9.24 \pm 4.84) \times 10^9\ M_\odot$ and the median HI mass below the median molecular gas mass is $M_{HI} = (2.22 \pm 0.56) \times 10^9\ M_\odot$. These calculations do not take the non-detections into account. The estimated values for the error bars are defined by determining the median deviation from the median values of the HI masses. Taking these error bars into account, it is found that the sources above the median molecular mass also have the larger HI masses, the sources below the median molecular gas mass have lower HI masses. Within the error bars the two populations of HI detected sources do not overlap. Nonetheless the HI subsample is very small and hence not suitable for making statistically significant statements.

4.5.1. Source confusion

The possibility of source confusion can not be excluded (see Fig. 4.1). The half power beam width (HPBW) at 21 cm is 9.5′. Since the whole beam lobe is (reduced) sensitive for radio signals out to twice the primary beam diameter, an HI rich galaxy at, the appropriate redshift, can cause disturbances out to angular distances of up to 18′. Asymmetries in the line profiles could possibly indicate the presence of companions or phases of merging that influence the host galaxies. One help to shed light on this matter may also be provided by the DSS images (see Fig. 4.1). Indications for interactions or companions can be looked for within a field of view representing the 9.5′ Effelsberg beam size. An outstanding example for this circumstance is the source HE 0150–0344. Its HI spectrum shows two emission lines close in velocity. In the DSS image the source shows an elongated structure with two peaks, indicating two very close by sources. Due to the large Effelsberg beam it is not possible to distinguish between the objects. Due to the close proximity of the sources interferometrical data would be needed to achieve the differentiation. Source confusion has a strong influence on the line shape, as well as on the line width, as mentioned earlier. Two objects, which exhibit emission at the observed frequency, close in redshift, hence also in recession velocity would cause a broadening of the studied emission line (always under the assumption that the two objects are galaxies). Therefore the target source would feature a much broader line width than it shows in reality (see Fig. 4.1, HE 0150–0344). This would then cause wrong numbers for the observed intensity and the resulting gas mass. Cross-checking with other observational tools, for e.g., optical images, is therefore essential to get the right results. Higher resolution spectroscopic observations would provide the actual redshift of the sources within the beam. This would enable the observer to differentiate between the objects in the given field of view.

4.5.2. Morphology

A comparison of the shape of the spectra shows good agreement between line shape and associated morphology in e.g., the optical (Fig. 4.1, Table 4.1, Table 4.2): For 11 of the 12 detected sources (the exception is HE 1248–1356) it was found that the morphology from the optical DSS images and the line shape determined from the HI spectra are the same. But one has to remark that the asymmetry of the line shape in some of the spectra makes it difficult to achieve an accurate differentiation. From the optical DSS images (Fig. 4.1) a spiral morphology was found in 15 (55%) host galaxies, elliptical morphology in 11 (41%) host galaxies and a ring morphology in 1 (4%) host galaxy. In 33% of the sources the HI spectra show triangular shaped line profiles, that is expected for turbulent line-of-sight dispersion (as in elliptical galaxies), and 67% have boxy or double-horn shaped line profiles, which is an indicator for emission from an inclined, rotating disk (as in spirals). In comparison Haan et al. (2008) find that 44% of the sources in their AGN sample are of spiral morphology, 28% are centrally peaked, 22% ringed and 5% of their host galaxies show a warped geometry. This hints at a preference for spirals in terms of morphological properties of QSO host galaxies. In their Seyfert subsample a slight excess of galaxies with a spiral geometry is evident.

In 1980 Dressler discovered a morphology-density (T-Σ) relation for galaxies: The fraction of elliptical like (E+S0) galaxies increases strongly with increasing local galaxy density, whereas the numbers of spiral galaxies decrease (see e.g., Table 1 in Pimbblet, 2003). In good agreement with further studies of this relationship, he also derives values for different environments: from galaxies in the field over poor and rich groups to clusters. Furthermore Dressler et al. (1997) report a redshift evolution in the T-Σ relation: an additional increase of the fraction of S0 galaxies with redshift whilst the number of elliptical galaxies stays nearly constant. The distribution of spiral, elliptical (E) and lenticular (S0) galaxies in the HI data is comparable to the numbers for field galaxies or poor groups. Dressler et al. (1985) expand the statistical studies of clusters by also looking at emission line galaxies. They find a much higher emission line frequency in field galaxies than in cluster galaxies. AGN occur at a rate of $\sim 5\%$ in field galaxies whereas in clusters only $\sim 1\%$ of the galaxies harbor an AGN. Martini et al. (2006) searched for low-luminosity AGN in low redshift clusters. They determined an AGN fraction of $\sim 5\%$, in clusters arguing that through X-ray emission more AGN towards lower luminosities can be identified. They concluded that if the AGN fraction in the field is indeed 5 times the AGN cluster fraction they would identify $\sim 25\%$ of the field galaxy population to show characteristics of AGN. On the other hand, Popesso & Biviano (2006) find an average AGN fraction of 18% in clusters. They, too, argue that their estimate includes low-luminosity AGN, probably missed in previous studies. Scaling the T-Σ relation from Dressler (1980) to AGN this would, by a strong simplification, mean that most AGN in the sample at hand reside in host galaxies of spiral morphological type, whereas in clusters the fraction of AGN in elliptical and lenticular hosts would increase significantly.

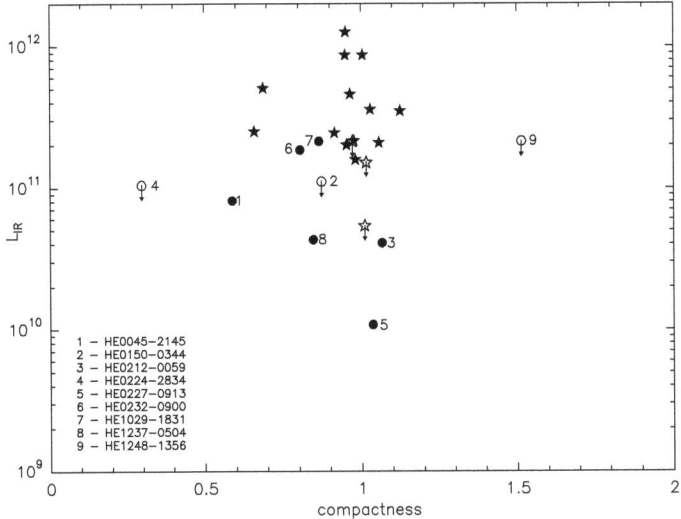

Figure 4.4.: Infrared luminosity L_{IR} vs. compactness f. Filled circles represent the HI detected sources from this work, unfilled ones with arrows mark upper limits of their infrared luminosities. Each detected object from this work is identified by a number, which is related to the source name given in the figure legend.

For analytical purposes an estimate for the compactness of gas confined to the host galaxies within the HI subsample by the derivation of the fraction of the CO line width to the HI line width is introduced as follows:

$$compactness f = \frac{v_{90}(CO)}{v_{90}(HI)}. \qquad (4.5)$$

v_{90} is the velocity width from which 90% of the emission arises. The values for the compactness are shown in Table 4.2. It was assumed that HI is found all over the place. If CO is mostly limited to the central inner region, filling the flat part of the rotation curve, f will be smaller than 1. For $f = 1$ the line widths in both emission lines should be equally large. Galaxies with compactness factors $f > 1$ have broader line widths in CO than in HI. As a large fraction of objects shows companions or signs of interactions, f may be used as an additional indicator for dynamical peculiarities (due to merging or disturbance). But as the beam size for the 21 cm observations was very large, other galaxies within the beam size contribute to the detected emission. Here the exceptional host galaxies HE 0232–0900 and HE 2302–0857, with their very broad emission lines at 21 cm have to be mentioned again. No clear trend could be seen in Fig. 4.4, where the IR luminosity vs. the compactness is shown. All sources from Haan et al. (2008) have compactness factors close to one, except for NGC 5953 and NGC 4321, which show factors $f < 1$ (Table 4.3). Usually a compactness factor close to one is expected as the CO distribution normally reaches the maximal rotation velocity. In this sense the considered sample is exceptional. For example the sources HE 0045–2145, HE 0224–2834, HE 0949–0122, HE 1248–1356 and HE 2302–0857 exhibit compactness factors strongly differing from a value of $f = 1$. But as the fraction of objects with companions and/or indications for mergers or interactions is extraordinary high in this source sample a compactness of $f < 1$, together with the line shape, may also provide early hints for distortions in the outer regions of the host galaxies. The best example for this issue is HE 0224–2834. The small compactness of 0.3 means that the atomic gas is much more extended than the molecular gas. The asymmetry in the shape of the HI line could imply that there are distortions in the outer HI disk. The comparison to the optical DSS images shows that indeed, HE 0224–2834 has a close companion and it shows morphological signs of interaction. The same holds true for HE 0045–2145: Its line shape is asymmetrical, the compactness factor is only 0.6 and in the DSS images there is another galaxy close by. Hence the HI distribution can serve as an early indicator for interactions. The compactness for objects with spiral features seems to be lower (the CO emitting regions have a smaller extension than HI emitting ones) as they have a larger outspread than the ones showing an elliptical geometry.

As another analytical tool the surface mass density was used:

$$\Sigma = \frac{M_{\mathrm{HI}}}{area} \; \left[M_\odot \, pc^{-2} \right] . \qquad (4.6)$$

The area of the sources again was estimated from the DSS images (Fig. 4.1, see section 4.4 again for the description of the area extraction process). The sample sources show areas between 55 and 76 000$''^2$ resulting in Σ values of 1.0 up to 434 $M_\odot \, pc^{-2}$. The median surface mass density, not taking

4.5. Results and discussion

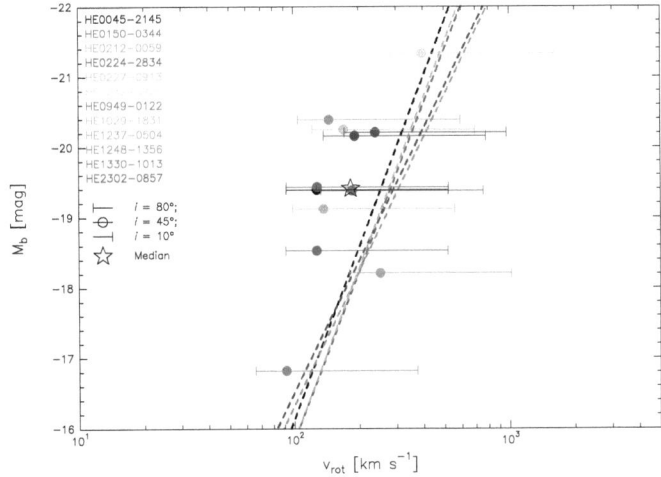

Figure 4.5.: Blue magnitude M_B vs. rotation velocity v_{rot}. The median value is plotted as an empty star. The dashed lines show functions for the Tully-Fisher relation from the literature (Schöniger & Sofue, 1997; Sakai et al., 2000; Ziegler et al., 2002). Each object is identified with its name given in the figure legend.

the non-detections into account, is of the order of 20.1 $M_\odot$ pc^{-2}, which is a factor of two lower compared to the value estimated for the Milky Way (50–75 $M_\odot$ pc^{-2}, Romano et al., 2000). The data from Haan et al. (2008) however result in an even lower average surface mass density of 1.77 $M_\odot$ pc^{-2}. Here the assumption of circular morphology leads to an overestimation of the area: Here the maximum HI radius was used and the assumption of a circular morphology of the host galaxy was made. Since the galaxies usually are not circular but more elliptical or exhibit other structures an assumption of the morphology was imprecise. This leads to significantly smaller surface mass densities for the sources than expected.

In Fig. 4.5, blue magnitude values M_b are plotted vs. the rotation velocity v_{rot}. This plot of the Tully-Fisher relation allows to test the quality of the FWHM values (W_{50}) for the emission lines. No severe outliers are seen that could not be explained as a consequence of the inclination. As the inclinations for the sources in this sample are not defined, only a range of rotation velocities can be assumed. By inserting graphs for the Tully-Fisher relation from literature

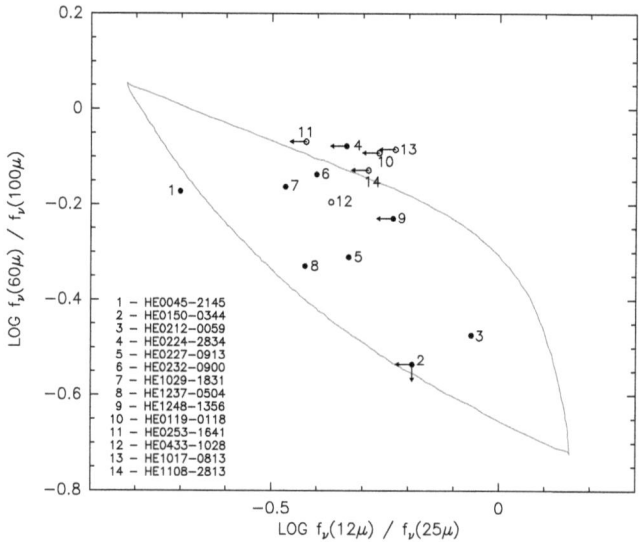

Figure 4.6.: IRAS color-color diagram according to Helou (1986). The underlying model was taken from Desert (1986). Filled circles show HI detected sources from this work, unfilled ones mark HI non-detected sources with available IRAS fluxes. Arrows are indicators for upper limits. Each object is identified by a number, which is related to the source name given in the figure legend. The locations of the HI sample sources in the plot show that almost all of them are starburst dominated hosts, showing high dust temperatures (sources to the upper left). The galaxies to the lower right of the plot have colors that are often found for elliptical galaxies. The two sources HE 0150–0344 and HE 0212–0059 are such objects. These objects are dominated by the 100 μm emission, implying lower dust temperatures.

sources (see e.g., Sakai et al., 2000; Ziegler et al., 2002), it can be estimated where the HI detected host galaxies from this sample should be found in the graph. Since the majority of the 45° points as well as the value for the median are located to the left of the literature Tully-Fisher functions, the objects from the HI subsample should feature lower inclination angles of about 30°, which is in good agreement with the assumption of a nearly face on inclination for low luminosity AGN.

4.5.3. IRAS color-color diagram

Figure 4.6 represents an IRAS color-color diagram of the sources from this sample after Helou (1986), overlaid with the computations of Desert (1986).

The HI subsample sources fit well into the model by Desert (1986). The location of the HI observed sources in the IRAS color-color plot shows that almost all of them exhibit signs which are typical for starburst dominated hosts that have elevated dust temperatures (sources to the upper left) i.e. show bright 60 μm emission compared to their 100 μm emission, and show faint 12 μm emission compared to their 25 μm emission. The far-IR color is particularly sensitive to foreground Galactic 'cirrus' emission, which can cause artificially small values of $f_\nu(60)/f_\nu(100)$. Only 2 objects (HE 0150–0344, HE 0212–0059) are dominated by the 100 μm emission implying lower dust temperatures. They have colors that are often found for elliptical galaxies. A visual inspection of their images shows that an identification with an elliptical host is plausible. Objects to the upper left of this plot are thought to be starburst galaxies.

4.5.4. Statistical considerations

4.5.4.1. Kendall-τ test

The Kendall-τ test provides a measure to determine the statistical significance of a statement obtained from a data set. It is the best, non-parametric method for testing for a correlation between two quantities with limits. The Kendall-τ test was used in order to do a 'proper' statistical test for a correlation between the atomic and molecular gas intensities.

The goal is to derive the value for τ from n data points (x_i, y_i). $\frac{1}{2}n(n-1)$ pairs of these data points are tested for a correlation between the rankings x and y. A pair of data points is called concordant if, for both n_i x and y show the same trend (i.e., $x_1 > x_2$ and $y_1 > y_2$). A pair of data points is called discordant if, for the n_i x and y show a different behavior (i.e., $x_1 > x_2$ but $y_1 < y_2$). In the case of $x_1 = x_2$ the pair is called t_y. If $y_1 = y_2$ the pair is called t_x. The results for all pairs of HI and CO intensities are shown in Table 4.4.

Table 4.4.: Kendall-τ test parameter overview

	1	2	3	4	5	6	7	8	9	10	11	12
HE 0045–2145 (1)	–	d[a]	c	d	d	c	d	c	d	c	d	c
HE 0150–0344 (2)	–	–	d	c	c	d	c	d	c	d	c	d
HE 0212–0059 (3)	–	–	–	d	d	c	d	c	d	c	d	c
HE 0224–2834 (4)	–	–	–	–	c	d	c	d	c	d	c	d
HE 0227–0913 (5)	–	–	–	–	–	d	c	d	c	d	c	d
HE 0232–0900 (6)	–	–	–	–	–	–	d	c	d	c	d	c
HE 0949–0122 (7)	–	–	–	–	–	–	–	d	c	d	c	d
HE 1029–1831 (8)	–	–	–	–	–	–	–	–	d	c	d	c
HE 1237–0504 (9)	–	–	–	–	–	–	–	–	–	d	c	d
HE 1248–1356 (10)	–	–	–	–	–	–	–	–	–	–	d	c
HE 1330–1013 (11)	–	–	–	–	–	–	–	–	–	–	–	d
HE 2302–0857 (12)	–	–	–	–	–	–	–	–	–	–	–	–

[a] A 'c' (for concordant) indicates that the HI and the CO intensity values follow the same trend for both observed objects, e.g. HE 0045–2145 and HE 0212–0059: For both sources their HI intensities have lower values than their CO intensities; 'd' indicates a disagreement.

Taking all the variables into account, τ can be derived with the following formula:

$$\tau = \frac{n_c - n_d}{\sqrt{n_c + n_d + t_y}\sqrt{n_c + n_d + t_x}} \quad (4.7)$$

n_c is the number of concordant pairs, n_d is the number of discordant pairs, t_x is the number of pairs where the x values are identical and t_y is the number of pairs where the y values are identic. For $\tau = +1$ the agreement between the two rankings, x_i and y_i, is perfect. If $\tau = -1$ the disagreement between the two rankings, x_i and y_i, is perfect. For τ values between -1 and +1 all other arrangements are represented. Increasing values imply an increasing agreement between the rankings. In the case of a total non-correlation (complete independence) between the rankings x and y (average for τ is 0), τ is an approximately normal distributed variable. Hence the variance s_τ is expressed as follows:

$$s_\tau = \frac{4n + 10}{9n(n-1)}. \quad (4.8)$$

The τ derived for the test on a correlation between the HI and CO intensities, in the HI subsample, resulted in a value of $\tau \sim -0.09$ with a variance s_τ of ~ 0.05. Hence, the Kendall-τ test shows that there is no correlation between the HI and CO intensities. But since the size of the sample is very small and therefore not at all statistically significant, a larger number of this kind of sources has to be observed in HI and CO emission.

4.6. Follow-up VLA observations

4.6.1. Observations and data reduction

In addition to the single dish 21 cm observations with the 100 m telescope in Effelsberg, VLA observations for two of the sample sources, HE 1248–1356, HE 1330–1013, were carried out in March 2008 at the same wavelength. The data were taken in the C configuration at an angular resolution of $13''$ and a primary beam size of $30'$. The 20 cm L-band receiver was used as the frontend, in conjunction with the VLA correlator as the backend. The correlator was used in spectral line mode with one IF (intermediate frequency) channel, 128 data channels, a bandwidth of 6.25 MHz and a channel separation of 48.828 kHz. The observations were centered on the redshifted rest frame frequency for each galaxy. The total integration time per source amounted to 6 hrs of which 3 hrs were spent on source. The data were reduced, imaged and analyzed with AIPS and IRAM's GILDAS software.

4.6.2. Results

Since at the time of the deadline for the VLA proposal only two of the sources of the whole HI subsample were detected successfully, only data for those two sources, HE 1246–1356 and HE 1330–1013, are available up to now. The HI continuum and line emission emission maps for HE 1246–1356 are presented in Fig. 4.7. The map shows that the HI emission measured with the 100 m Effelsberg telescope is indeed coming from the position of HE 1248–1356. The value for the neutral atomic gas mass determined from the interferometric map results in $\sim 1.0 \times 10^9 \, M_\odot$, which is in good agreement with the value determined from the HI single-dish observations.

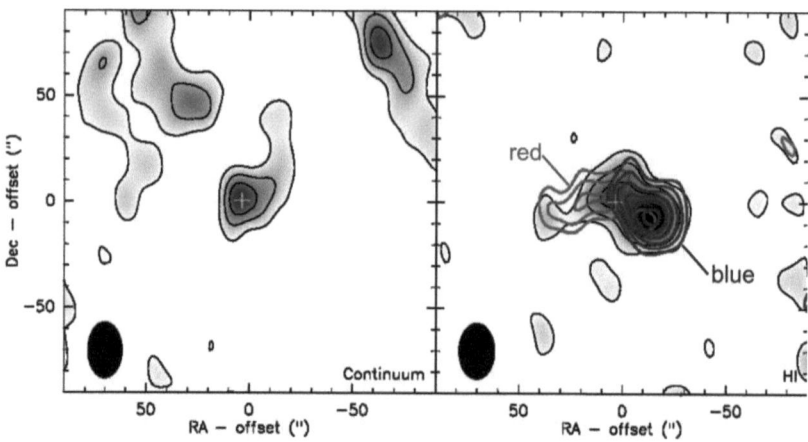

Figure 4.7.: Left: Continuum map (3σ contour steps starting at the 3σ level). **Right:** HI map (black contours, 2 and 1σ contour steps for the line emission, starting at the 2σ level, and the red- and blueshifted gas, starting at the 3σ level). The cross marks the position pointed at in the observations. The grey contours to the left of the source position represent the redshifted HI gas and the grey contours to the right represent the blueshifted gas.

For HE 1330–1013 no 21 cm HI line emission could be found at the position of the QSO. The HI emission detected with the telescope in Effelsberg seems to originate from another source within the primary beam size (30′) of the VLA. The only properties known so far for this source, NVSS J133205–101949 (with a RA(J2000) of 13:32:05.0 and a DEC(J2000) of -10:19:49.7), are its position from the NVSS catalog (*NRAO VLA Sky Survey* Condon et al., 1998) and the intensity of the source in the NVSS continuum observations at 20 cm (49.1±2.3 mJy at 1.4 GHz). Most likely, the HI emission line detected in Effelsberg was inserted via the side lobes of the telescope's power diagram.

4.7. Conclusions

In this chapter of the thesis the results on the neutral atomic gas content of 27 nearby low-luminosity QSO host galaxies were presented. 12 of the target sources were detected in HI emission. The atomic gas masses cover a range from 1.1×10^9 $M_\odot$ up to 3.8×10^{10} $M_\odot$ with a median mass of the order of 6.2×10^9 $M_\odot$. Considering the upper limits for non-detections this value changes to 11.4×10^9 $M_\odot$. Taking this median HI mass for the whole HI subsample into account, the atomic hydrogen mass is a factor of two higher for this works sample sources than the HI mass of the Milky Way ($M_{HI} = 5.5 \times 10^9$ $M_\odot$, see Hartmann & Burton, 1997). Though it is not possible to discuss the nature of the objects without high resolution imaging in more detail, it was possible to draw some interesting conclusions with the help of the profiles of the HI spectra and optical DSS images. For several galaxies, possible companions in the vicinity were found, as well as indications for merger/interactions. No distinct correlation between the IR luminosity of a galaxy and its HI gas content was seen. In general galaxies with high IR luminosities have high neutral atomic hydrogen masses. The compactness for objects with spiral features seems to be lower (the CO emitting regions have a smaller extension than HI emitting ones) as they have a larger outspread than the ones showing an elliptical geometry. Only for the sources from the sample at hand there are CO and HI measurements done for the sample together. For other samples, like e.g., for the Haan et al. sample, there exist CO measurements with different telecopes/techniques in the literature by separate authors. To my knowledge an estimated number of about ~ 50 sources have HI and CO data taken. Therefore the HI subsample contributes to the, currently, rather small contingency of hosts with atomic and molecular masses. The observed median surface mass density is about a factor two lower than the values for the Milky Way (50–75 $M_\odot$ pc^{-2}). From the Tully-Fisher relation it was deduced that a median inclination of $\sim 30°$ is in good agreement with the assumption of nearly face-on morphology. Interferometrical data through follow-up observations for two out of the twelve HI detected sources have been obtained. For the other ten sources these follow-up observations are in planning.

5. A search for H$_2$O maser emission in nearby low-luminosity QSO host galaxies

5.1. Introduction

Water vapor masers are excellent tracers of physical conditions, such as temperature ($T_{\rm kin} > 400$ K) and density ($n(H_2) > 10^7$ cm^{-3}) of the molecular gas, in the highly obscured innermost parts of active galactic nuclei. Extragalactic water masers with isotropic luminosities $L_{\rm H_2O} > 10\ L_\odot$ are classified as *megamasers*. But since the apparent luminosity is derived under the assumption of isotropic emission, the true luminosity may be smaller by several orders of magnitude (see e.g., the recent review by Lo, 2005). Henkel et al. (2005) found a transition near $L_{\rm H_2O} = 10\ L_\odot$ between weaker masers mostly related to star formation and stronger masers associated with active galactic nuclei. Furthermore, they found a correlation between the luminosity in the infrared and the total isotropic water maser luminosity.

Towards Seyfert and LINER galaxies, H$_2$O megamasers are used to probe the small scale structure and kinematics of accretion disks or tori, to obtain the masses of the central engines through a determination of Keplerian rotation, and to derive geometric distances to the parent galaxies (e.g., Miyoshi et al., 1995; Herrnstein et al., 1999). Complementing these so-called *disk-masers* are the *jet-masers*, which are either arising from the interaction of the nuclear jet with an encroaching molecular cloud, or from a cloud being accidently located along the line-of-sight to the jet. In one source, Circinus, a nuclear *outflow-maser* was also detected. To date, H$_2$O vapor megamasers are found in 10% of observed AGNs in the local universe (Braatz et al., 2004). They are almost exclusively found in Seyfert 2 and LINER type galaxies (Braatz et al., 1997; Kondratko et al., 2006b), i.e. mostly in spirals. This can be interpreted in terms of the unified scheme, indicating that in Seyfert 2 galaxies the nuclear disk is seen roughly edge-on, and that LINER galaxies are also hosting an AGN. Because the nuclear disks of spirals are not aligned with the large scale morphology of the parent galaxy, a convincing correlation between the inclination of the large scale disks and the presence of megamasers is not apparent.

The first quasar discovered to emit water megamaser emission at 22 GHz, (J080430.99+360718.1, Barvainis & Antonucci, 2005), was classified as a type 2 quasar. The fact that the second H_2O megamaser emitting quasar is a type 1 quasar (MG J0414+0534, Impellizzeri et al., 2008) brings the (statistically non-significant) detection ratio for quasars to 1:1. Since QSOs have more massive cores than Seyfert galaxies, the chance to find maser emission in type 1 or type 2 quasars could totally differ from what is known for Seyfert galaxies. So the question is: Could the detection by Impellizzeri et al. (2008) just be a serendipitous one, or could this detection of water megamaser emission in a type 1 quasar be trend-setting for the local universe? In order to find an answer to this question, the sample at hand was searched for water masers in southernly nearby low-luminosity QSO host galaxies. In Sect. 5.2 the sample is described, Sect. 5.3 is devoted to the characterization of the observations and Sect. 5.4 comprises the results and discussion. Sect. 5.5 gives a short summary.

5.2. The sample

As a continuation of the study of low-luminosity nearby QSO host galaxies, sources from the same volume limited parent sample as in Chapter 4 were observed. Now, in this chapter of the thesis the 17 most luminous, at infrared wavelengths, sources of the original CO subsample (Table. 1), were observed in the 22.235 GHz H_2O vapor maser transition. They feature recession velocities ranging from 7 900 km s^{-1} to 18 200 km s^{-1}. In addition to the 17 nearby QSO hosts, two galactic sources, known to show water maser emission, were observed as 'control sources' and calibrators: W3(OH) and Orion–KL.

5.3. Observations and data reduction

The observations were carried out in the 6_{16}–5_{23} 22.235 GHz H$_2$O maser transition on 27 and 28 November 2007 using the Effelsberg 100-m telescope. The 18–26 GHz two-channel K-band HEMT facility receiver was used as the frontend in conjunction with a 8 192-Channel-Autocorrelator (AK 90) and a Fast Fourier Transform Spectrometer (FFTS) as backends. The latter provided a bandwidth of 500 MHz with a channel width of ~ 0.4 km s^{-1}, thus covering a velocity range of approximately 7 150 km s^{-1}. The AK90 in NSPLIT mode 25 consisted of eight channel backends with a somewhat broader channel width of ~ 1 km s^{-1} and an individual bandwidth of 40 MHz. Some of the eight backends were shifted in frequency to cover altogether a total bandwidth of 80 MHz. The FFTS was used in the load switching mode, employing the rotating horn of the 1.3 cm primary focus receiver that guarantees excellent baselines, with 3 minutes on and 3 minutes off the source positions. The rotating horn was switched between two fixed positions with a frequency of 1 Hz and a beam throw of 120″. The beam efficiency was ~ 0.53 for a beam size of 40.2″. Pointing checks, using sources from the Effelsberg Catalog of pointing and flux density calibration (Ott et al., 1994), were used at least once every hour. For the flux calibration the continuum cross scans of NGC 7027 were used. NGC 7027 has a continuum flux density of 5.9 Jy (taken from Baars et al., 1977; Mauersberger et al., 1987).

Data reduction and analysis were performed using the GILDAS CLASS package. If applicable all spectra have been averaged. To all spectra a baseline has been fitted, which was subtracted subsequently. In addition, each subscan was corrected for the elevation dependency of the gain individually. The in-

5. A search for H$_2$O maser emission in nearby low-luminosity QSO host galaxies

Table 5.1.: Sources searched for H$_2$O maser emission

Object	RA(2000) [h] [m] [s]	DEC(2000) [°] [′] [″]	v_0 (LSR) [kms^{-1}]	z	D_L [Mpc]	ν [MHz]	$t_{obs,on}$[a] [min]	rms [Jy]	Intensity I[b] [Jy kms^{-1}]	Seyfert[c] Type	Morphological[d] Type	M_{BH}[e] [$M_\odot$]
HE0021-1819	00:23:55.3	-18:02:50	15954	0.053	215.4	21111.93	27	0.013	< 0.53 ± 0.16	–	E, C	–
HE0040-1105	00:42:36.8	-10:49:21	12578	0.042	169.6	21338.85	27	0.013	< 0.29 ± 0.16	Sy 1.5	E, C	$10^{6.70}$
HE0114-0015	01:17:03.6	+00:00:27	13682	0.046	184.3	21265.38	27	0.011	< 0.02 ± 0.14	–	E, C, poss. M	$10^{6.80}$
HE0119-0118	01:21:59.8	-01:02:25	16412	0.055	221.5	21081.90	27	0.010	< 0.39 ± 0.13	Sy 1.5	–	–
HE0150-0344	01:53:01.4	-03:29:24	14329	0.048	193.3	21220.73	27	0.011	< 0.39 ± 0.14	–	E	–
HE0212-0059	02:14:33.6	-00:46:00	7921	0.026	106.2	21663.17	27	0.014	< 0.10 ± 0.18	Sy 1.2	E, C	$10^{7.20}$
HE0224-2834	02:26:25.7	-28:20:59	18150	0.060	245.4	20966.60	27	0.020	< 0.40 ± 0.25	–	M	–
HE0232-0900	02:34:37.7	-08:47:16	12886	0.043	173.7	21318.39	27	0.015	< 0.24 ± 0.19	Sy 1	M, C	–
HE0345+0056	03:47:40.2	+01:05:14	8994	0.031	124.9	21566.52	27	0.016	< 0.37 ± 0.20	Sy 1	R, M, C	$10^{8.05}$
HE0433-1028	04:36:22.2	-10:22:33	10651	0.036	143.1	21472.80	54	0.011	< 0.49 ± 0.14	Sy 1	E, C	–
HE0853-0126	08:56:17.8	-01:38:07	17899	0.060	242.1	20982.43	54	0.008	< 0.76 ± 0.10	Sy 1	S	–
HE1011-0403	10:14:20.6	-04:18:41	17572	0.059	237.6	21004.23	27	0.011	< 0.28 ± 0.14	Sy 1	S, C	$10^{7.03}$
HE1017-0305	10:19:32.9	-03:20:15	14737	0.049	199.0	21192.41	93	0.007	< 0.06 ± 0.09	Sy 1	S, C, poss. M	–
HE1107-0813	11:09:48.5	-08:30:15	17481	0.058	236.3	21010.19	27	0.012	< 1.02 ± 0.15	Sy 1	E	–
HE1126-0407	11:29:16.6	-04:24:08	18006	0.060	243.7	20974.51	27	0.012	< 0.10 ± 0.15	Sy 1	S, C	–
HE2233+0124	22:35:41.9	+01:39:33	16913	0.056	228.5	21047.97	45	0.009	< 0.20 ± 0.11	Sy 1	S, C	–
HE2302-0857	23:04:43.4	-08:41:09	14120	0.047	190.4	21234.92	30	0.012	< 0.12 ± 0.15	Sy 1.5	S, C	$10^{8.54}$
W3(OH)	02:27:04.1	+61:52:22	-46.4[f]	0.000	0.6	22235.08	12	0.087	9944 ± 0.59	–	–	–
Orion-KL	05:35:14.2	-05:22:22	7.74[g]	0.000	0.1	22235.08	18	0.040	19203 ± 0.32	–	–	–
resampled[h]	–	–	–	–	–	–	–	0.001	< 0.07 ± 0.05	–	–	–

[a] total observation time spent on the source

[b] Upper limits and errors of the intensity represent 1 σ values.

[c] taken from the NED

[d] E denotes elliptical morphology; S represents a spiral morphology; M stands for possible mergers or merger remnants; R denotes ringed objects; C marks galaxies with other extragalactic sources within a projected distance of up to 340 kpc

[e] References: for HE0040-1105, HE0114-0015: Greene & Ho (2006a); for HE0212-0059: Greene & Ho (2006b); for HE0232-0900, HE2302-0857: O'Neill et al. (2005); for HE1011-0403: Wang & Lu (2001)

[f] Velocity taken from Bronfman et al. (1996).

[g] Velocity taken from Matveenko et al. (2000).

[h] All maser spectra of the nearby QSO sample sources were resampled and averaged to obtain these values.

88

tensity errors ΔI (see Table 5.1 for the results) were determined following the procedure from Bertram et al. (2007). The geometric average of the line error $\Delta I_\mathrm{L} = \sigma\, v_\mathrm{res}\, \sqrt{N_\mathrm{L}}$ and the baseline error $\Delta I_\mathrm{B} = \sigma\, v_\mathrm{res}\, N_\mathrm{L}/\sqrt{N_\mathrm{B}}$ were taken into account. σ is the rms noise in Jy, v_res the spectral resolution in km s^{-1}, N_L the number of channels over which the line spreads and N_B the number of channels used for fitting a polynomial to the baseline.

5.4. Results/Discussion

The observation of 17 objects in the water maser subsample of low-luminosity QSO host galaxies resulted in no H_2O detections at individual 3σ levels of 27 up to 60 mJy (Table 1, Appendix A). Smoothing the data in order to obtain a better signal-to-noise ratio didn't improve the situation.

To prove that the method, applied for the observations, worked, two known galactic maser sources, namely Orion–KL and W3(OH) (see Table 5.1), were observed. W3(OH) was primarily chosen as a source for the flux calibration. The observed spectrum for Orion–KL is shown in Fig. 5.1.

Following the procedure of Goldsmith et al. (2008) it was assumed that all host galaxies do exhibit H_2O megamaser emission at a weak level. To allow for different maser emission velocities in different sources, the following procedure was adopted: At first a section of the spectrum, around the velocity of the CO emission, was selected for each host. As the offset between the central velocity and the individual components of the megamaser emission can easily be about several hundreds of km s^{-1} (for examples see e.g., Kondratko et al., 2006a; Greenhill, 2007; Braatz & Gugliucci, 2008), the whole spectrum was taken into account. Each spectrum is then cross-correlated with a Gaussian whose width represents the expected line width of the maser emission. In the case of a well-known line width of the megamaser emission each spectrum would then be shifted by the offset between the central channel and the channel of maximum cross-correlation amplitude. To achieve the adjustment to the same velocity offset for all sources the reference channel of

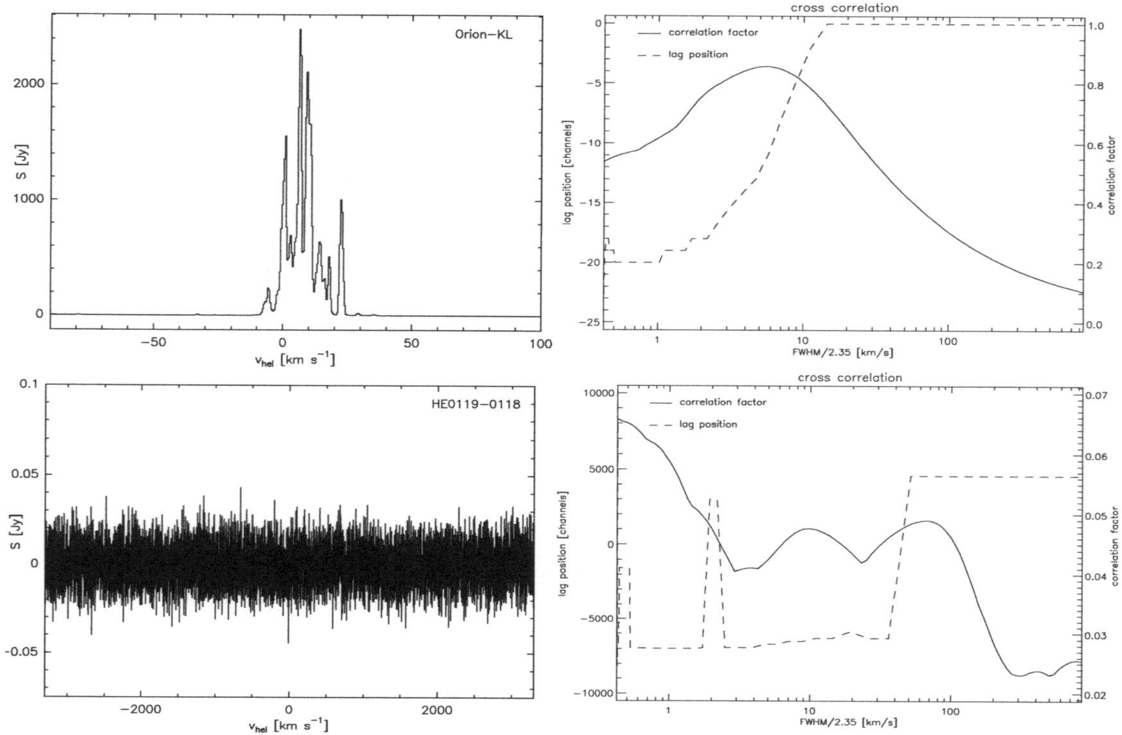

5. A search for H$_2$O maser emission in nearby low-luminosity QSO host galaxies

Figure 5.1.: Top left: Observed spectrum of the galactic known maser source Orion–KL at 22 GHz. **Bottom left:** Observed spectrum of the nearby QSO HE 0119–0118 at 22 GHz. This spectrum is shown as a representative for the H$_2$O maser spectra of all observed nearby QSO host galaxies. **Right:** Results of the cross-correlations between the 22 GHz spectra of the two sources, Orion–KL (top) and HE 0119–0118 (bottom), and Gaussian line profiles of different line widths.

the central velocity (formerly defined by the heliocentric velocity of the observed galaxy) in each spectrum has been artificially set to zero. Following this procedure, the shifted data for all observed sources were added together. The described procedure will create a line feature resembling the correlation template (a Gaussian signal in this case) from the constructive alignment of purely random noise. But since the line width of water megamasers can differ between few km s^{-1} (NGC 520: FWHM = 1.1 km s^{-1}; Castangia et al., 2008) over several tens (NGC 2989: FWZI = 30 km s^{-1}; Braatz & Gugliucci, 2008) up to even some hundred km s^{-1} (UGC 3193: FWZI $\sim$ 350 km s^{-1}; Braatz & Gugliucci, 2008) it was decided to do a study of the cross-correlation parameters first.

Each spectrum was correlated with several Gaussian profiles. The line widths of the Gaussians range from the value of the spectral resolution for each individual spectrum up to 2 000 km s^{-1}. From the cross-correlation the correlation factor and the lag between the peak positions of the Gaussian and the spectrum itself were derived. In order to compare the so gained values to sources known to exhibit maser emission, a correlation on the two galactic objects Orion–KL and W3(OH) was performed. For a plot of the results on Orion–KL see Fig. 5.1. As a representative member of the QSO hosts see the plot on the results of the cross-correlation on HE 0119–0118 (Fig. 5.1).

In Fig. 5.1 the correlation factor clearly peaks at a FWHM/2.35 of 5.35 km s^{-1} which corresponds to a FWHM of 12.60 km s^{-1} of the Gaussian profile. The lag is the number of channels by which the velocity corresponding to the central reference channel has to be shifted left or right in order to match the reference channel to the channel of maximum cross-correlation amplitude. As the correlation coefficient for Orion–KL has a maximum (0.856) at a lag of -18 channels the Gaussian has to be shifted by 18 channels to the right hand side of the velocity reference channel in the spectrum to get the maximum agreement between Gaussian profile and spectrum.

Correlation factor and lag position in Fig. 5.1 however show a different behavior. The curve for the correlation factor peaks at three different FWHM. Furthermore the factor ranges between factors of only 0.02 and 0.07 indicating the absence of a correlation. Beyond that the lag position oscillates between shifts of −7 000 and +5 000 channels. The same behavior can be seen in the plots for the cross-correlation of the other QSO host galaxies (see Appendix A).

Although the procedure of the cross-correlation described above could be a measure to distinguish between detections and non-detections, it has some caveats. First of all, the analysis of data with this method may only be appropriate for observations with a good signal-to-noise ratio. Secondly, the FWHM of the line should be known up to some extent. Furthermore one should always keep an eye on the correlation factor and the development of the lag. Too small correlation factors are not trustable in terms of the confirmation of a possible detection. If the spread of the lag oscillation grows too wide (see e.g., Fig. 5.1) the central line position is not stable, hence indicating either several line components or signals from random noise or radio frequency interference.

5.4.1. Sensitivity

As Bennert et al. (2009) already pointed out, surveys for H_2O megamaser emission put high requirements on the conditions during the observations. Since the emission is usually very weak ($\ll 1$ Jy), a high sensitivity is required. The widths of maser lines can be of the order of km s^{-1}, or even below (≤ 1 km s^{-1}), but also of the order of hundreds of km s^{-1} (e.g., UGC 3193; Braatz & Gugliucci, 2008). This factor makes a high spectral resolution during the observations a prerequisite. Since the offset between the central velocity of the observed object and the individual components of the megamaser emission can easily be separated by about several hundreds of km s^{-1}, the bandwidth is another delimiting factor which plays a big role in the observations of water megamaser emission lines. Additionally, the unknown radial velocities of the maser components make a large bandwidth inalienable. With the velocity coverage of $\sim 7\,200$ km s^{-1} enough band width (effective 420 MHz with the FFTS) was provided for the observations to cover a large velocity range in order not to miss red- or blue-shifted maser components.

5.4. Results/Discussion

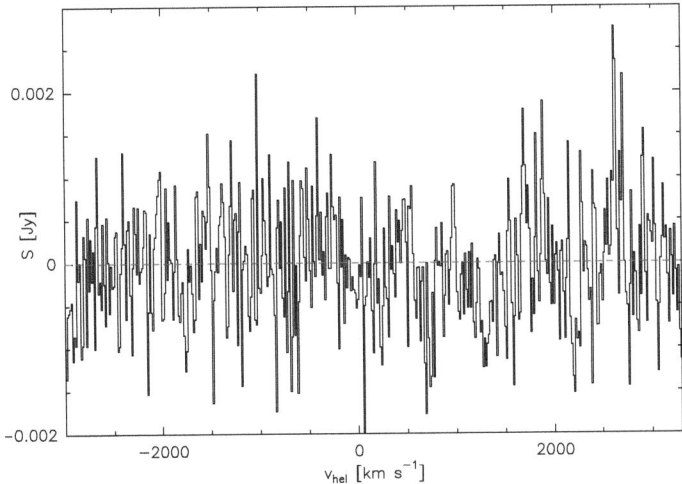

Figure 5.2.: Smoothed spectrum of the resampled and averaged observations of all nearby QSO host galaxies. Before averaging, the spectra were resampled to the same velocity resolution of 0.4367 km s^{-1} and the central velocities were shifted to zero. The channel spacing of the displayed spectrum is ~ 14 km s^{-1}. For the averaging process the spectrum of each observed source was weighted equally.

No trend could be seen for one prominent FWHM for the Gaussian line profiles in the cross-correlation plots. Hence, the lag shifts were not executed but rather all spectra were simply added together, after the resampling to the same spectral resolution. This assumes that the emission occurs at the same velocity offset with respect to the velocity of the central channel in any given spectrum. The resulting resampled and averaged spectrum is shown in Fig. 5.2.

The limits on the maser intensity were determined under the assumption of a FWZI (full width at zero intensity) of 350 km s^{-1}, which is one of the values for the broadest water maser lines (UGC 3193; Braatz & Gugliucci, 2008). For the H$_2$O subsample sources an average rms noise value of 12 mJy was obtained. In comparison to this, values for samples in the literature were looked at where water megamaser have been detected. Braatz et al. (1996) state a sensitivity of 60 mJy in ~ 1 km s^{-1} channels. Greenhill et al. (2003) and Kondratko et al. (2006b) observed at a rms of 14 mJy in 1.3 km s^{-1} channels. In 2006, Kondratko et al. published a second paper stating an average rms noise in a 24.4 kHz channel of 4.6 mJy. Braatz & Gugliucci (2008) and

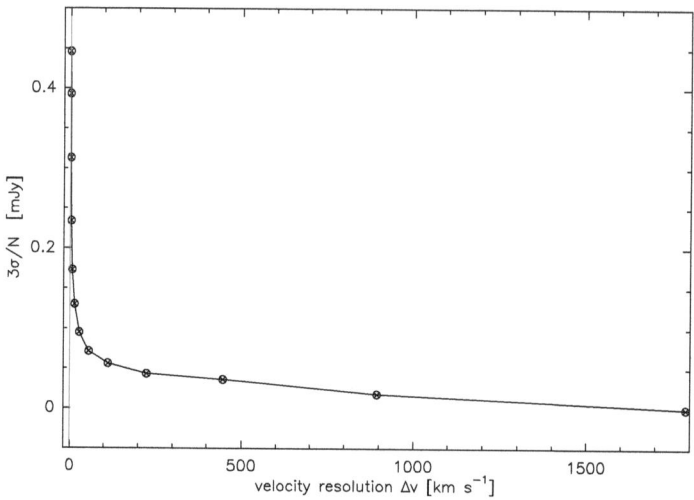

Figure 5.3.: 3 σ-over-N (number of observed sources in the sample) as a function of the spatial resolution serving as a mean upper limit of the emission in each galaxy.

Castangia et al. (2008) published their observations with rms noise levels of 6 mJy per 0.33 km s^{-1} channel and an average rms of 14.33 mJy. In this timely order of observations throughout the years there is the clear tendency that the sensitivity increases with time significantly. The average rms value of 12 mJy is placed well within the range of sensitivities where detections in theory should be possible.

In order to get a handle on the upper limit of the emission from the observed frequencies at 22 GHz, the 3 σ-*over*-N value for the resampled spectrum was determined for 13 different velocity resolutions (Fig. 5.3). σ is the channel-to-channel rms noise and N is the number of host galaxies observed in the H$_2$O subsample. Fig. 5.3 shows that with growing channel width the 3 σ/N value goes down, which means that by accumulating signal through the binning of several channels the signal-to-noise ratio increases. Since in this chapter of the thesis only non-detections of H$_2$O megamaser emission are reported, this plot is used as a measure to get an estimation of a mean upper limit of the emission in each sample galaxy.

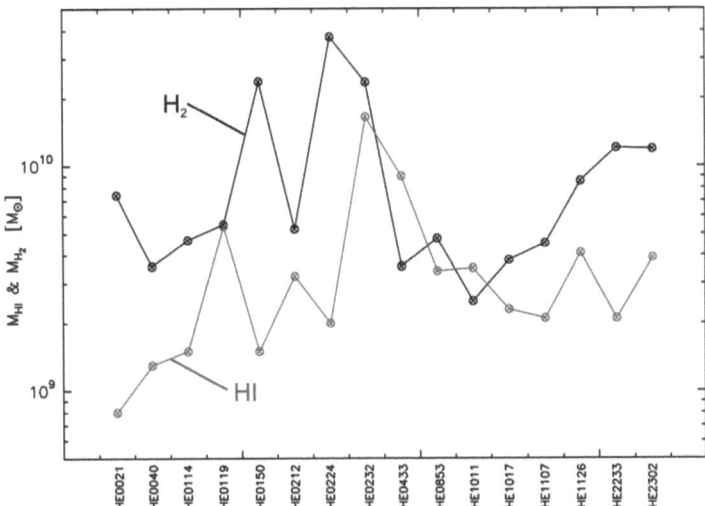

Figure 5.4.: Comparison of the atomic and molecular gas masses of the sources searched for water maser emission. Values for HI masses are depicted by black circles, grey circles represent the H_2 masses derived from CO observations.

5.4.2. Host galaxy properties

16 of the 17 observed sources were detected in molecular gas emission (CO (2−1) & CO(1−0), Bertram et al., 2007). In addition, five of the CO detected sources which were searched for H_2O megamaser emission have been detected previously in 21 cm HI line emission (Chap. 4). Fig 5.4 shows the neutral atomic HI gas mass M_{HI} (values from Chap. 4) and the molecular gas mass M_{H_2} derived from CO observations (values from Bertram et al., 2007). It is surprising to see that for 14 of the 16 nearby QSO hosts with the highest luminosities in the IR, the HI mass is larger than the molecular gas mass. Nonetheless, this seems to be only a 'local' trend in this sample. To achieve statistical significance on this subject the sample considered is too small.

5.4.3. Morphology

In 78 galaxies of different types, water megamasers have been detected so far (Bennert et al., 2009). The water megamaser observed at the highest redshift was detected in MG J0414+0534 at z = 2.639 (Impellizzeri et al., 2008). The largest part in the population of masing galaxies is made up by Seyfert types (78%). 11% are LINERs (low ionization nuclear emission regions) and 7% are HII galaxies. The smallest percentage of galaxies showing water megamaser signatures are starburst galaxies (3%) and NLRGs (narrow-line radio galaxies). Although Seyfert galaxies present the majority of maser galaxies only 3% are Seyfert 1s, whereas Seyfert 2s are the dominant type (88%). In Quasars H_2O megamaser emission was already detected not only in a type 2 quasar (Barvainis & Antonucci, 2005), but also in a type 1 quasar (Impellizzeri et al., 2008), which, furthermore, is the megamaser with the highest redshift.

Braatz et al. (1997) found that the absence of detections in Seyfert 1 galaxies indicates either that these galaxies do not have molecular gas with appropriate conditions to mase, or that the masers in these galaxies are beamed away from the observers line of sight. The latter is in good agreement with the findings of Braatz & Gugliucci (2008), who state that masers, specifically those in AGN accretion disks are beamed in the plane of the disk. Miyoshi et al. (1995) show with VLBA observations of NGC 4258 that masers are found in thin edge-on disks only few parsec away from the supermassive black hole in the nucleus. The unified scheme of AGN (Lawrence & Elvis, 1982; Antonucci, 1993) says that Sy 1 nuclei are hidden within Sy 2 galaxies behind an obscuring dusty molecular thick disk, or torus, indicating that the two different types of Seyfert galaxies only differ in terms of the viewing angle: Sy 1s are seen more pole-on while Sy 2s are seen more edge-on. If one takes the detection rate of H_2O megamasers in all AGN (10%) into account, the probability to find no maser emission is at 90%. It was tried to determine the viewing angle under which the probability to find megamaser emission is 10%, which is the fraction of AGN known to exhibit water maser emission.

$$P(\vartheta) = 2 \cdot \frac{2\pi(1 - \cos\vartheta)}{4\pi} = 1 - \cos\vartheta \qquad (5.1)$$

Putting the values into Eq. 5.1 results in an angle of 6°. This means that the maser emission most probably will be detected if the line of sight to the

5.4. Results/Discussion

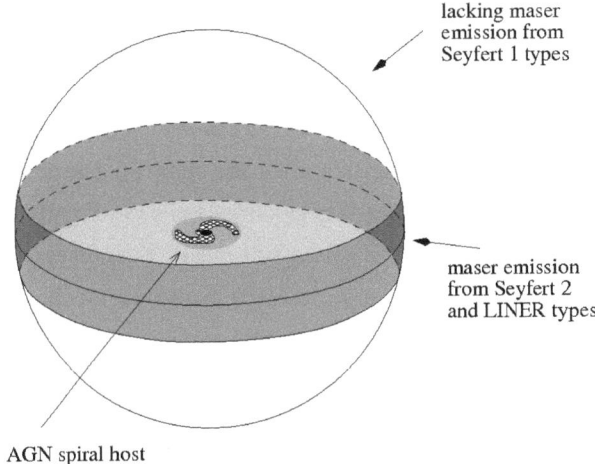

Figure 5.5.: A simple depiction of the unified scheme in context of water megamaser emission. Sy 2 galaxies are the host galaxies with the most water maser detections. A possible explanation in terms of the unified scheme is that the only difference between Sy 1 galaxies and Sy 2s is the viewing angle of the observer. Sy 2 galaxies are seen more edge-on whereas Sy 1s are seen more pole-on. The angle under which the maser emission is detectable is only ± 6° from the equatorial plane (region shaded in grey). This explains the lack of megamaser detections in Sy 1 galaxies.

observer falls within an angular distance of ± 6° from the equatorial plane. For that see also Fig. 5.5. This statement is in good agreement with the findings of Miyoshi et al. (1995). Nuclei with large hydrogen column densities ($N_{H_2} > 10^{24}$ cm^{-2}) seem to favor the excitation of H$_2$O masers (Braatz et al., 1997; Kondratko et al., 2006b; Madejski et al., 2006; Zhang et al., 2006). This accentuates the importance of the obscuring torus geometry additionally. In this context Miyoshi et al. (1995) furthermore find that if other H$_2$O megamasers have a similar structure to the ones found in NGC 4258, the detection of only type 2 Seyferts indicates that the thick torus and the thin disk tend to be coplanar. As shown by Greenhill et al. (1996) (high-resolution observations of NGC 1068), the maser emission seems to arise from the innermost parsecs of the nucleus or rather the inner edge of the thick torus. Braatz et al. (1997) suggest that the scale height of the inner thin disk might expand with the increase of the distance from the black hole and by that present a transition to the thick disk or torus obscuring the Sy 1 nucleus. Furthermore, they asses that the absence of maser detections in Sy 1 galaxies cannot be hold responsible for the weak tendency of H_2O-detected galaxies to be intrinsically fainter than undetected ones.

In their statistical overview of all known megamaser sources Bennert et al. (2009) find that most galaxies known to host a water megamaser are classified as spirals (84%). The remaining 16% are made up by S0 galaxies (7%), elliptical galaxies (1%) and irregular or peculiar galaxies (8%). Only NGC 1052, which is classified as a Seyfert 2 galaxy or LINER (Bennert et al., 2009) has an elliptical morphology. They argue that the mechanism fueling the nuclear activity is one important property separating spirals (morphological type of most Seyferts) from early type galaxies (morphological type of most QSOs). Bar instabilities may be the means to transfer the gas, provided by the spiral host galaxy, into the central region to fuel the AGN hosted by Seyfert galaxies (Combes, 2006). The H_2O subsample consists of eight elliptical or S0 host galaxies, six spirals and three galaxies that seem to be in a merger/interaction phase, which unfortunately decreases the, a priori already small, chance to find a water maser in a sample composed of Sy 1–1.5, even more. Taking the statement of Bennert et al. (2009) about the morphology of the galaxies known to show maser emission into account the chance of finding one maser from an elliptical host galaxy is very slim. Statistically, one out of 78 detected water megamaser sources shows elliptical morphology, which means it would take $\sim 1\,900$ observed galaxies to find a second elliptical host galaxy with water megamaser emission. In addition, the continuum fluxes are very low for the sample sources. A strong continuum could amplify the 22 GHz H_2O maser emission. Jets are not known for any of the sample galaxies.

5.4.4. Black hole mass

One prominent difference between Seyfert galaxies and QSOs, besides the brightness on the absolute magnitude scale, lies in the mass of their nuclear engines. Seyferts have black hole masses between 10^6 and few $10^7\, M_\odot$ (e.g., Herrnstein et al., 1999; Henkel et al., 2002). Quasars, on the other, hand can reach masses of the central black hole up to $10^9\, M_\odot$ (e.g., Labita et al.,

2006; Vestergaard et al., 2008). The black hole masses for the sources in this works sample range from rather small, at least for QSOs, $1.07 \times 10^7\ M_\odot$ (HE 1011–0403; Wang & Lu, 2001) up to large $3.47 \times 10^8\ M_\odot$ (HE 2302–0857; O'Neill et al., 2005). The large black hole masses could possibly imply that the conditions in the vicinity of the nuclear black hole are not stable enough to provide the conditions necessary for stable megamaser emission (Tarchi et al., 2007; Bennert et al., 2009).

5.5. Conclusions/Summary

Here the results on a search for H_2O water megamasers in 17 nearby low-luminosity QSO host galaxies are presented. None of the target sources was detected in the 22 GHz maser emission line. Therefore the results of previous water megamaser surveys, stating that extragalactic water maser are found primarily in Seyfert 2 galaxies and LINERs, are supported. The atomic and molecular gas contents of the member galaxies of the H_2O subsample were compared. For almost all of them the atomic gas content is larger than the molecular one. A sensitivity study shows that the observational setup used obtaining the discussed data was sufficiently suited to detect water megamaser emission. To prove this in practise two known galactic maser sources (Orion–KL and W3(OH)) were observed successfully.

A. Appendix

A.1. The H$_2$O maser spectra and DSS2 IR images

A. Appendix

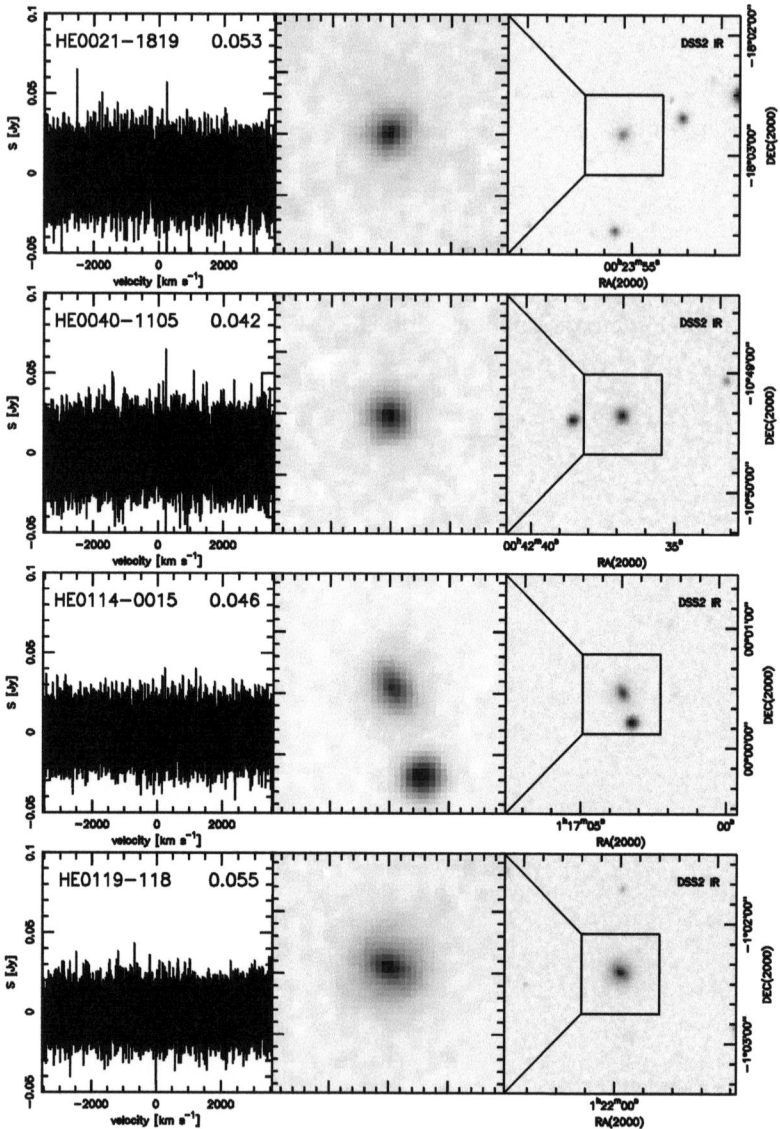

Figure A.1.: 22 GHz maser spectra of the observed host galaxies from the 'nearby QSO sample' observed with the Effelsberg 100-m telescope and optical DSS images all objects. The images in the middle extend over 40.2″ and the ones to the right contain 2′. 40.2″ is roughly the size of the beam at 22 GHz. North is up and East to the left. Each source is identified by its HES name (top left corner of the spectrum) and the redshift (top right corner).

A.1. The H$_2$O maser spectra and DSS2 IR images

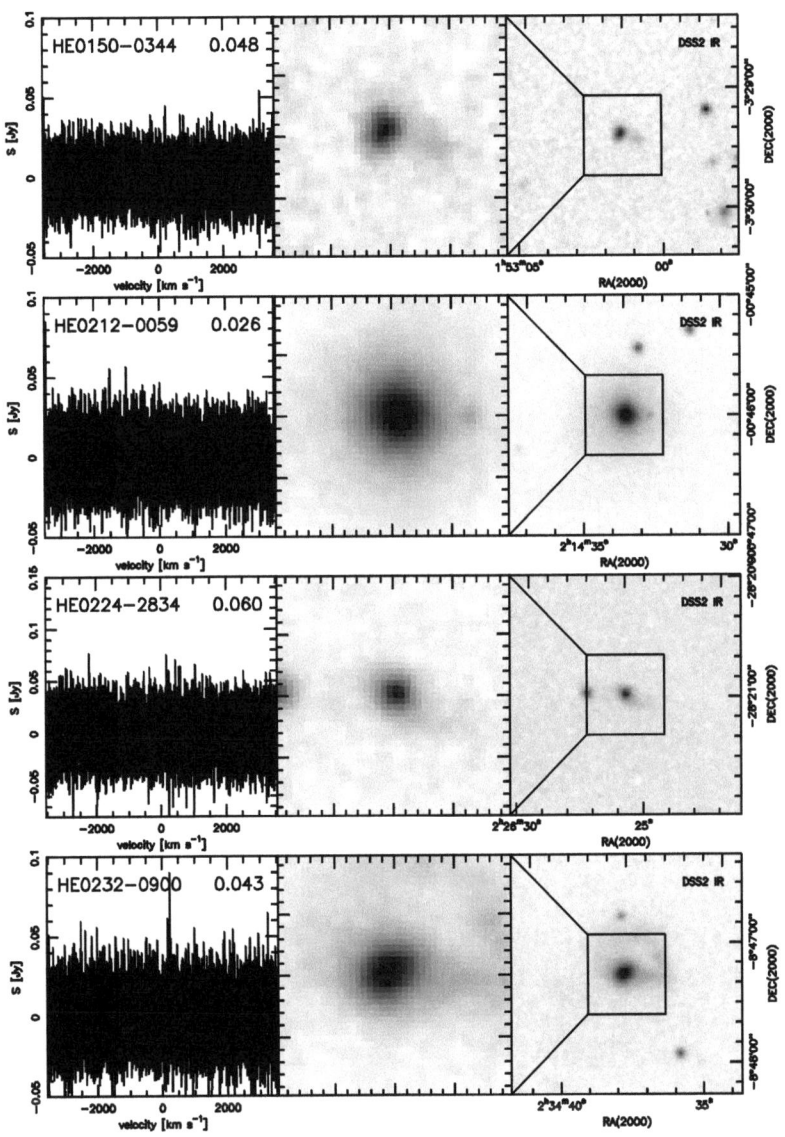

Figure A.1.: continued

A. Appendix

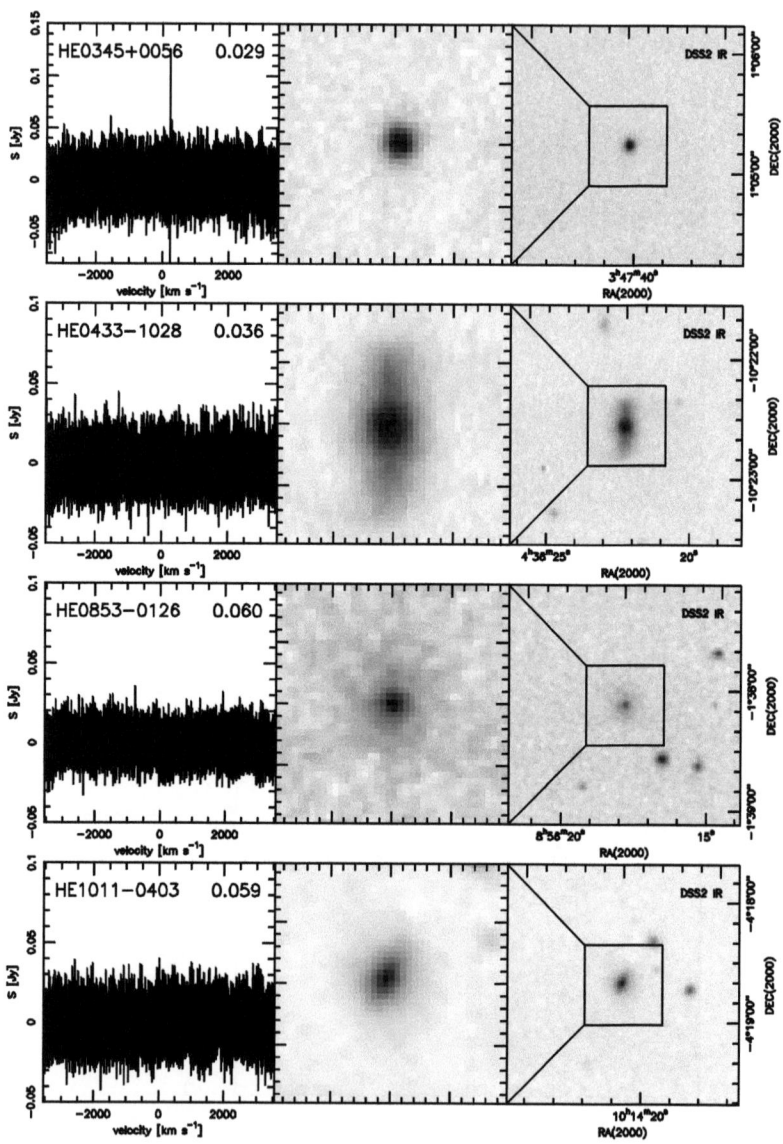

Figure A.1.: continued

A.1. The H$_2$O maser spectra and DSS2 IR images

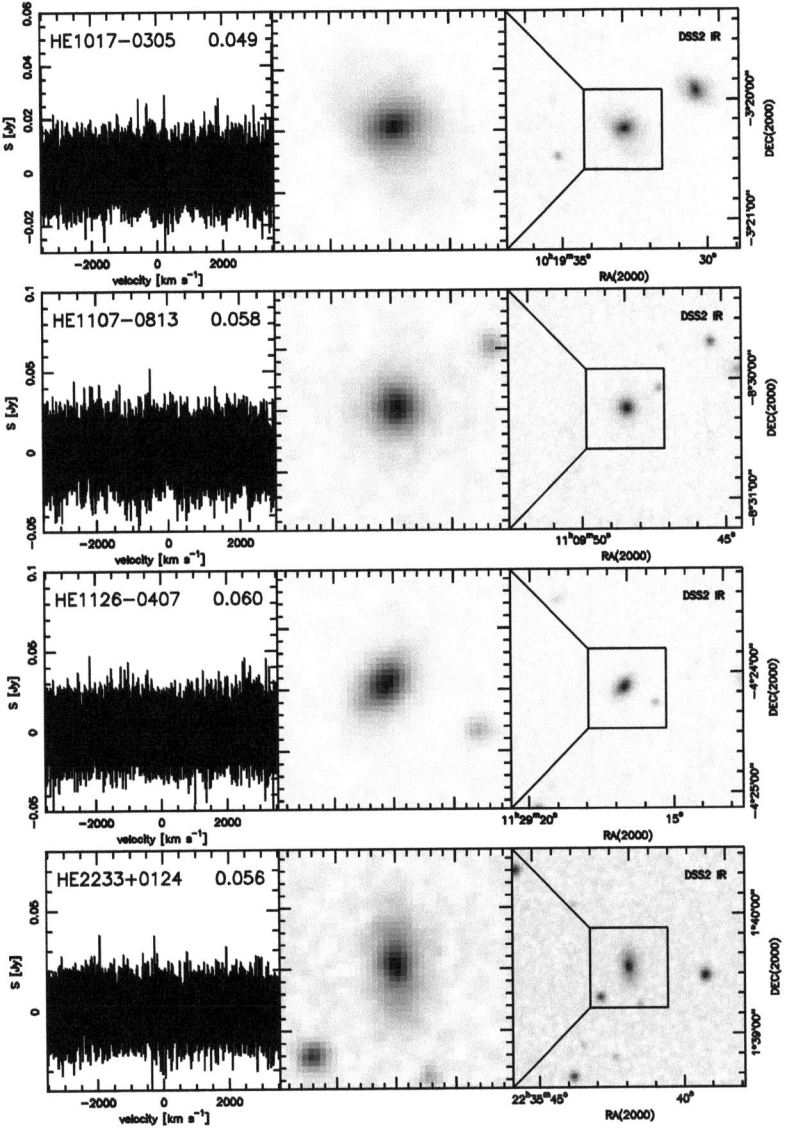

Figure A.1.: continued

A. Appendix

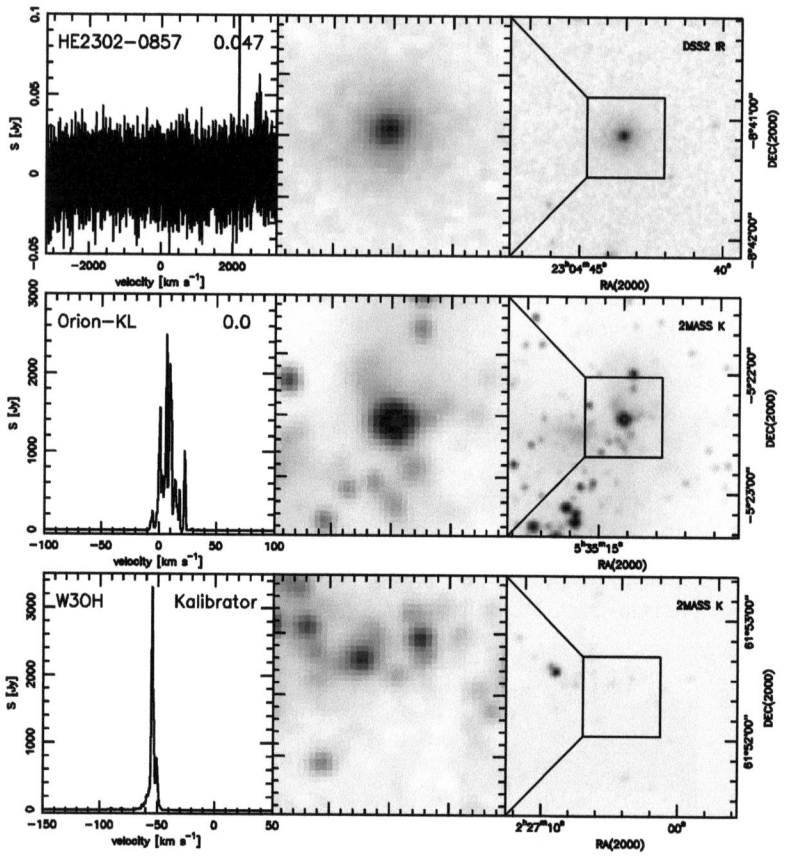

Figure A.1.: continued

A.2. The H$_2$O maser spectra and the corresponding cross-correlation plots

A. Appendix

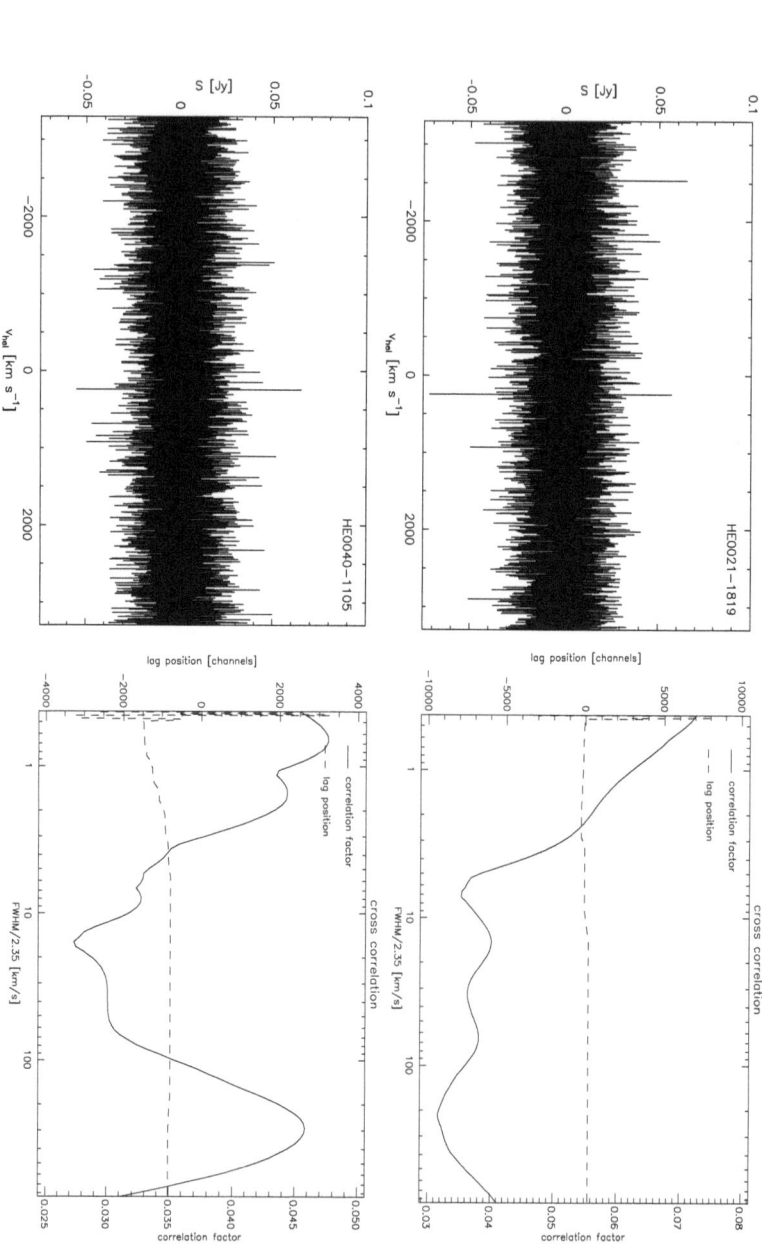

Figure A.2: **Left:** Observed spectrum of the nearby QSOs HE 0021−1819 (*top*) and HE 0040−1105 (*bottom*) at 22 GHz. **Right:** Results of the cross-correlations between the 22 GHz spectra of the two sources, HE 0021−1819 and HE 0040−1105, and Gaussian line profiles of different line widths.

A.2. The H₂O maser spectra and the corresponding cross-correlation plots

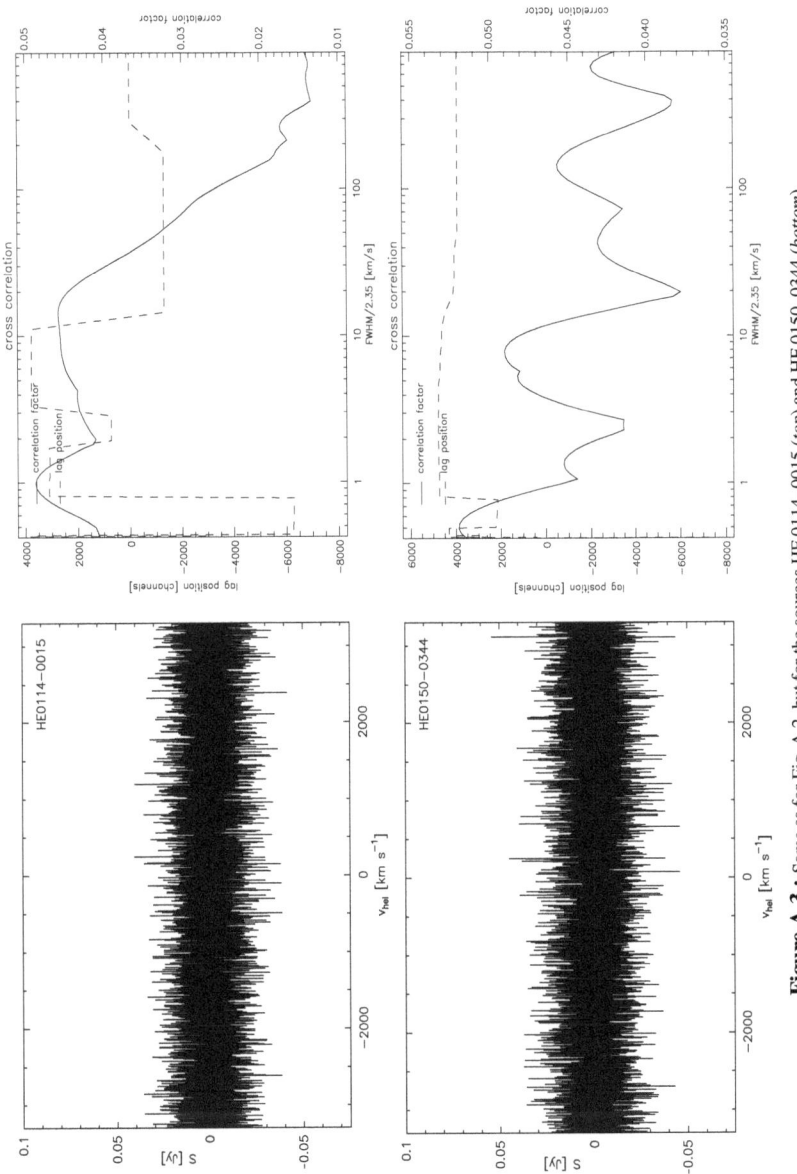

Figure A.3.: Same as for Fig. A.2, but for the sources HE 0114−0015 (*top*) and HE 0150−0344 (*bottom*).

A. Appendix

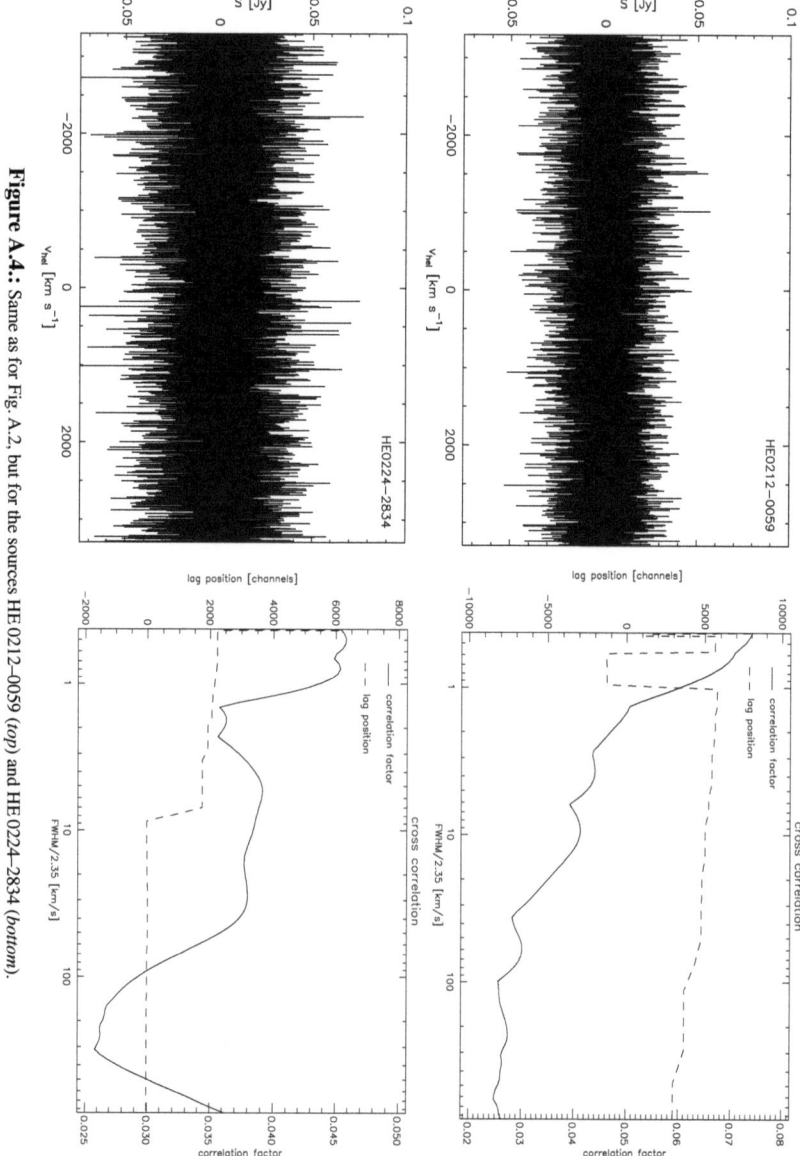

Figure A.4.: Same as for Fig. A.2, but for the sources HE 0212−0059 (*top*) and HE 0224−2834 (*bottom*).

A.2. The H$_2$O maser spectra and the corresponding cross-correlation plots

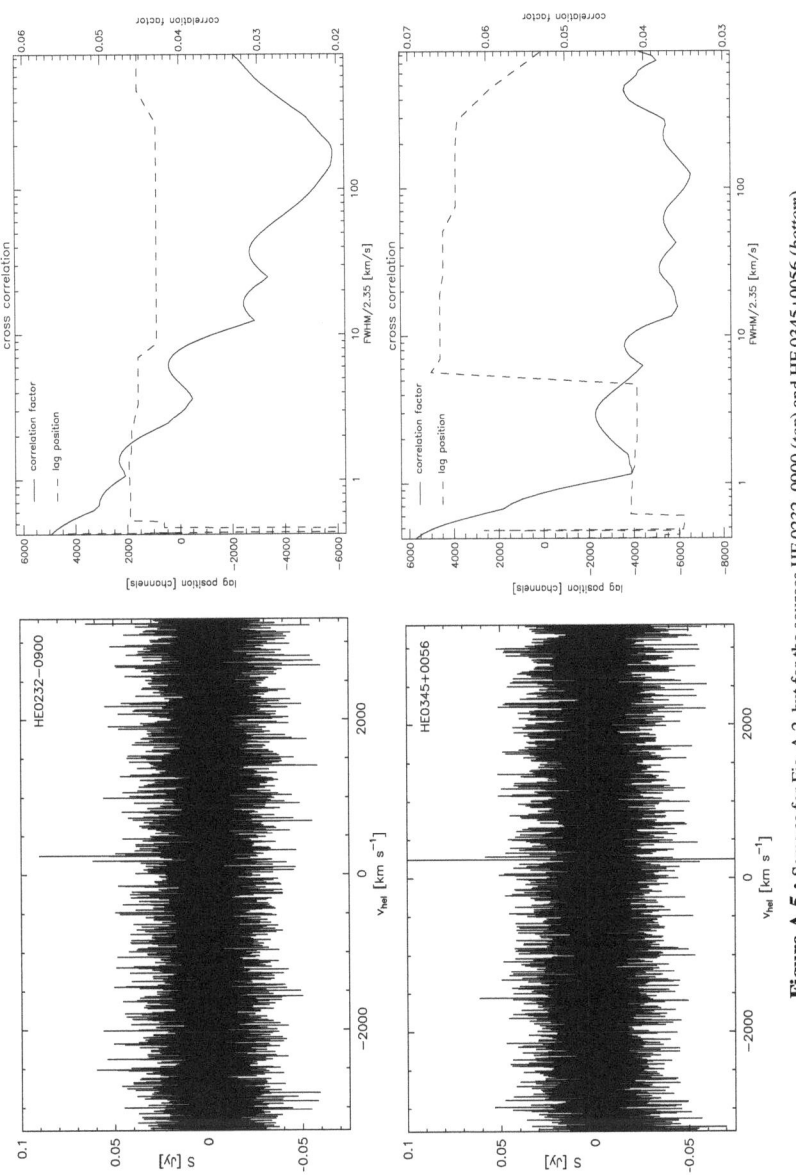

Figure A.5.: Same as for Fig. A.2, but for the sources HE 0232−0900 (*top*) and HE 0345+0056 (*bottom*).

A. Appendix

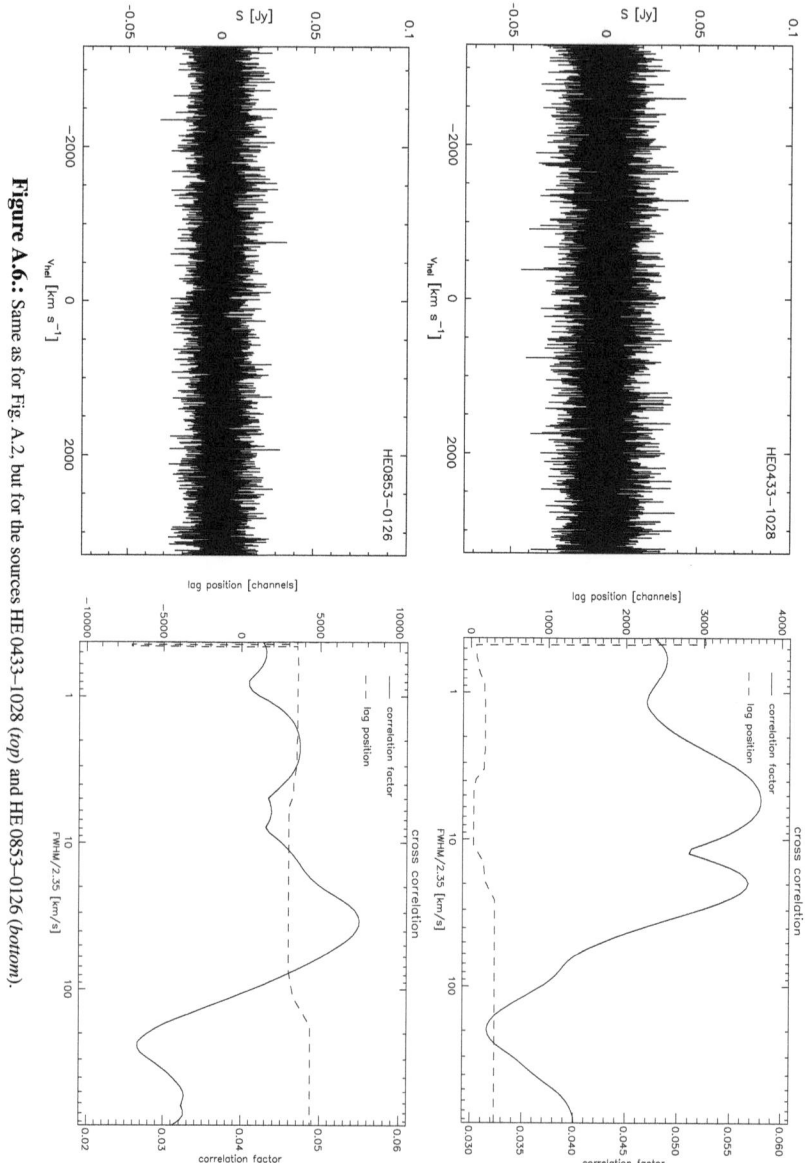

Figure A.6.: Same as for Fig. A.2, but for the sources HE 0433−1028 (*top*) and HE 0853−0126 (*bottom*).

A.2. The H$_2$O maser spectra and the corresponding cross-correlation plots

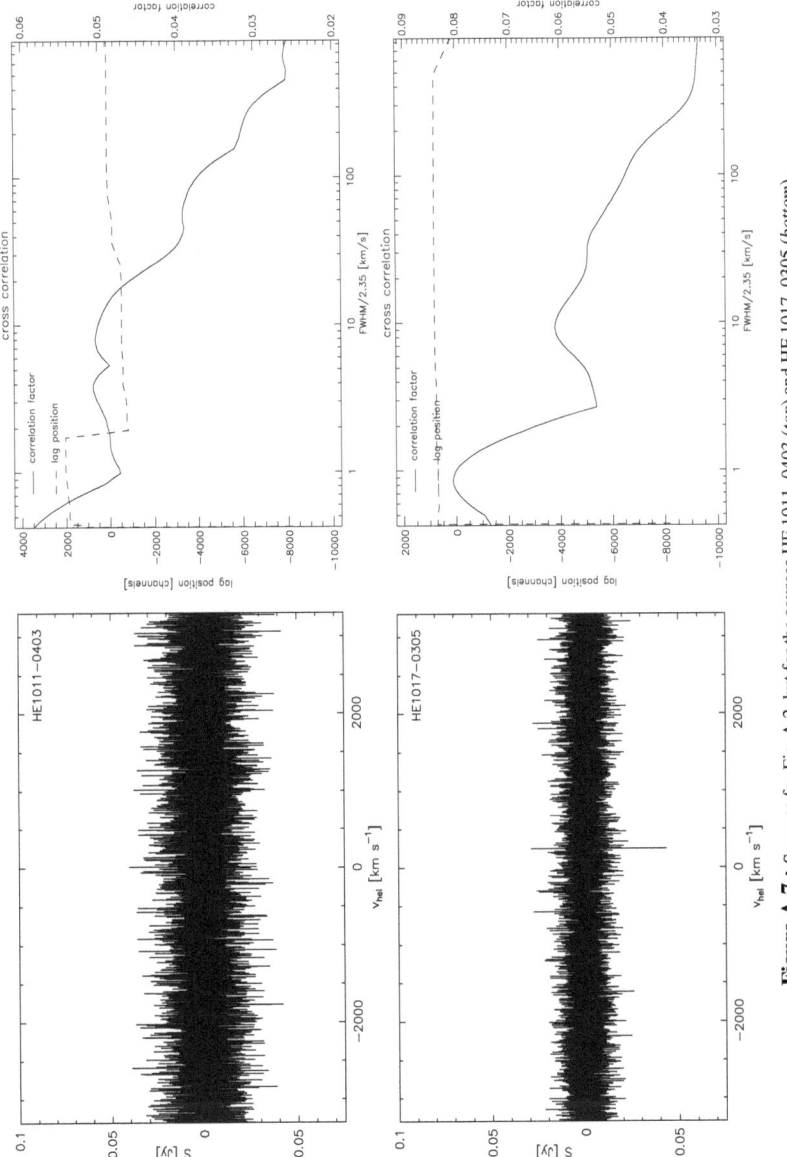

Figure A.7.: Same as for Fig. A.2, but for the sources HE 1011−0403 (*top*) and HE 1017−0305 (*bottom*).

A. Appendix

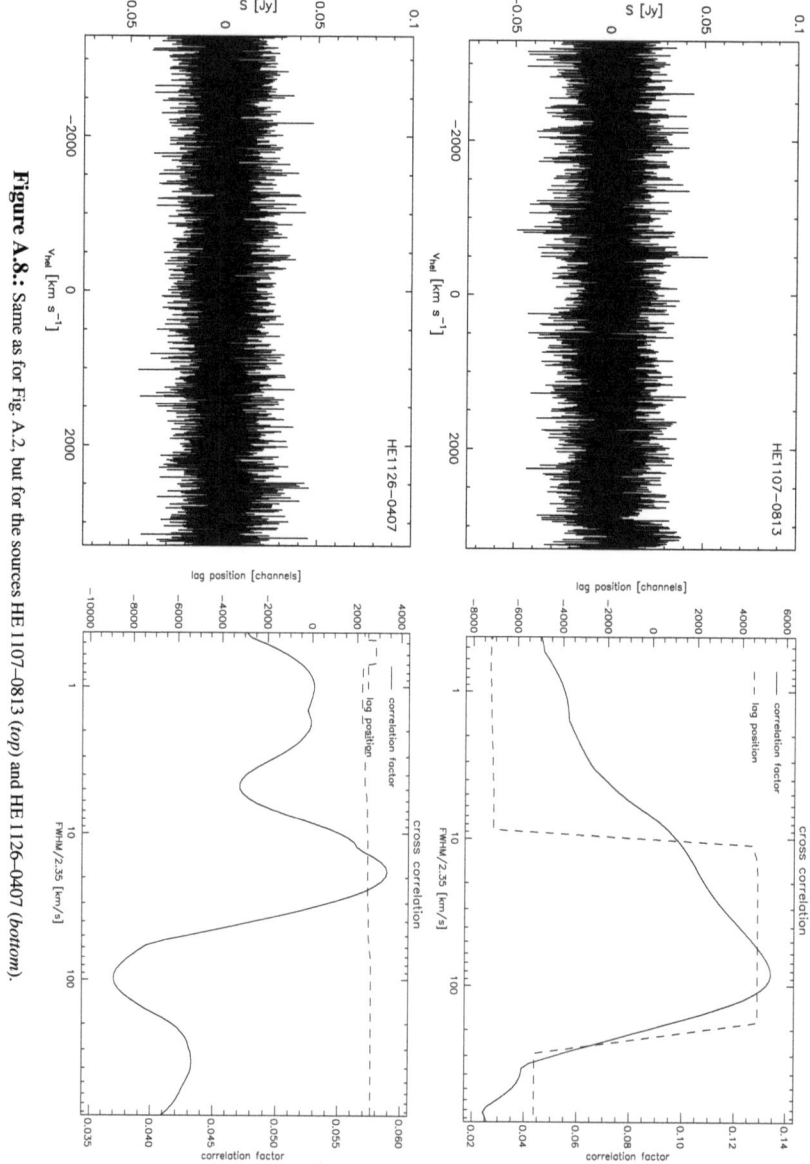

Figure A.8.: Same as for Fig. A.2, but for the sources HE 1107−0813 (*top*) and HE 1126−0407 (*bottom*).

A.2. The H$_2$O maser spectra and the corresponding cross-correlation plots

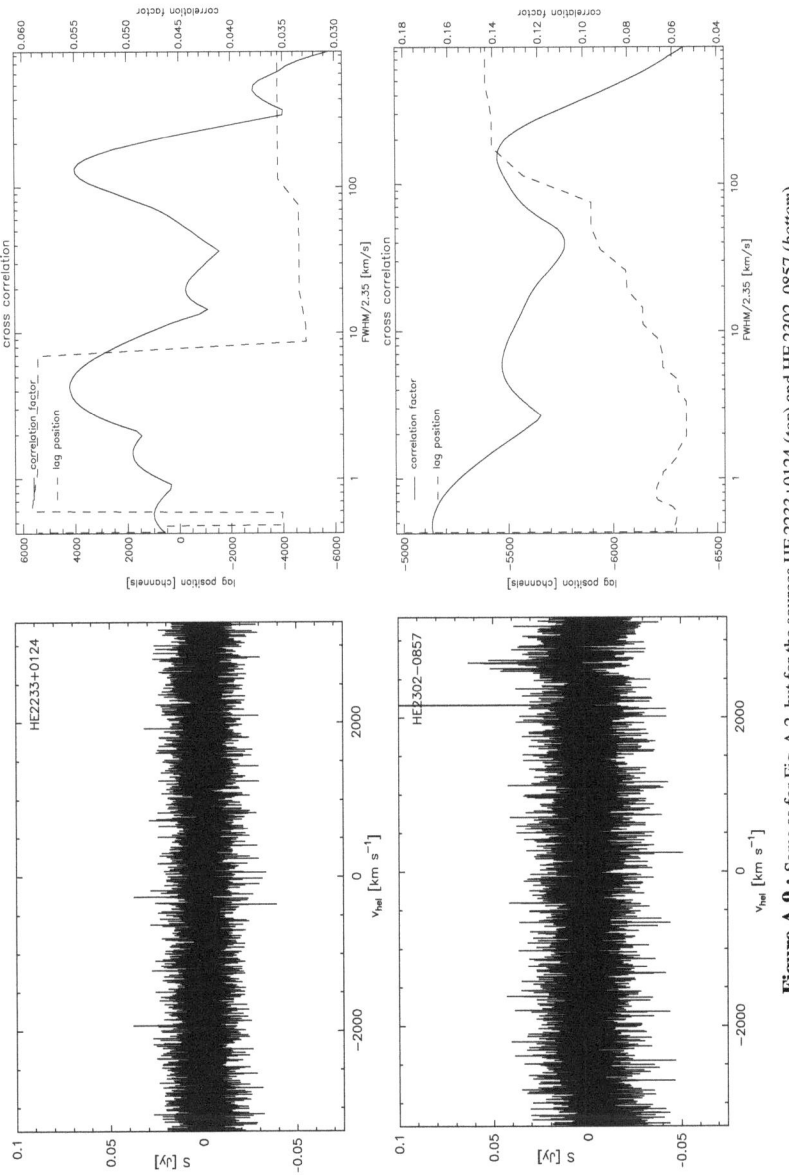

Figure A.9.: Same as for Fig. A.2, but for the sources HE 2233+0124 (*top*) and HE 2302−0857 (*bottom*).

A. Appendix

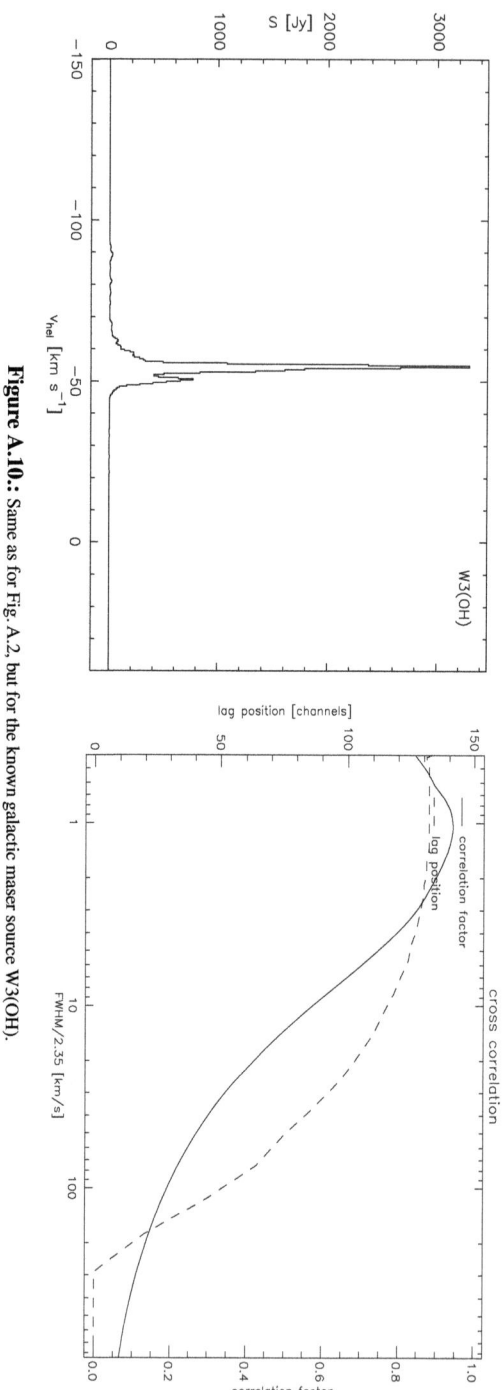

Figure A.10.: Same as for Fig. A.2, but for the known galactic maser source W3(OH).

Bibliography

Aalto, S., Johansson, L. E. B., Booth, R. S., & Black, J. H. (1991). Large CO-12/CO-13 intensity ratios in luminous mergers. *A&A*, 249, 323–326. 70

Antonucci, R. (1993). Unified models for active galactic nuclei and quasars. *ARA&A*, 31, 473–521. 9, 96

Arp, H. (1966). Atlas of Peculiar Galaxies. *ApJS*, 14, 1–+. 35

Arribas, S., Colina, L., & Clements, D. (2001). Two-dimensional Kinematical and Ionization Structure of the Warm Gas in the Nuclear Regions of Arp 220. *ApJ*, 560, 160–167. 42

Baan, W. A. (2007). Arp 220 IC 4553/4: understanding the system and diagnosing the ISM. In J. M. Chapman & W. A. Baan (Eds.), *IAU Symposium*, volume 242 of *IAU Symposium* (pp. 437–445). 35

Baars, J. W. M., Genzel, R., Pauliny-Toth, I. I. K., & Witzel, A. (1977). The absolute spectrum of CAS A - an accurate flux density scale and a set of secondary calibrators. *A&A*, 61, 99–106. 87

Bahcall, J. N., Kirhakos, S., Saxe, D. H., & Schneider, D. P. (1997). Hubble Space Telescope Images of a Sample of 20 Nearby Luminous Quasars. *ApJ*, 479, 642–+. 10, 57

Bajaja, E., Huchtmeier, W. K., & Klein, U. (1994). The extended HI halo in NGC 4449. *A&A*, 285, 385–388. 17

Barnes, J. E. (1988). Encounters of disk/halo galaxies. *ApJ*, 331, 699–717. 46

Barnes, J. E. & Hibbard, J. E. (2009). Identikit 1: A Modeling Tool for Interacting Disk Galaxies. *AJ*, 137, 3071–3090. 45, 46

Barvainis, R. & Antonucci, R. (2005). Extremely Luminous Water Vapor Emission from a Type 2 Quasar at Redshift z = 0.66. *ApJ*, 628, L89–L91. 86, 96

Baudry, A. & Brouillet, N. (1996). Spatial distribution and nature of H_2O masers in Messier 82. *A&A*, 316, 188–195. 21

Bennert, N., Barvainis, R., Henkel, C., & Antonucci, R. (2009). A Search for H_2O Megamasers in High-z Type-2 Active Galactic Nuclei. *ApJ*, 695, 276–286. 92, 95, 98, 99

Bertram, T., Eckart, A., Fischer, S., Zuther, J., Straubmeier, C., Wisotzki, L., & Krips, M. (2007). Molecular gas in nearby low-luminosity QSO host galaxies. *A&A*, 470, 571–583. 19, 20, 58, 60, 64, 67, 69, 71, 72, 89, 95

Binney, J. & Merrifield, M. (1998). *Galactic astronomy.* 8

Boselli, A., Casoli, F., & Lequeux, J. (1995). CO observations of spiral galaxies in the Virgo cluster and in the Coma/A1367 supercluster. *A&AS*, 110, 521–+. 70

Boyce, P. J., Disney, M. J., Blades, J. C., Boksenberg, A., Crane, P., Deharveng, J. M., Macchetto, F. D., Mackay, C. D., & Sparks, W. B. (1998). HST Planetary Camera images of quasar host galaxies. *MNRAS*, 298, 121–130. 57

Braatz, J. A. & Gugliucci, N. E. (2008). The Discovery of Water Maser Emission from Eight Nearby Galaxies. *ApJ*, 678, 96–101. 89, 91, 92, 93, 96

Braatz, J. A., Henkel, C., Greenhill, L. J., Moran, J. M., & Wilson, A. S. (2004). A Green Bank Telescope Search for Water Masers in Nearby Active Galactic Nuclei. *ApJ*, 617, L29–L32. 85

Braatz, J. A., Wilson, A. S., & Henkel, C. (1996). A Survey for H_2O Megamasers in Active Galactic Nuclei. I. Observations. *ApJS*, 106, 51–+. 93

Braatz, J. A., Wilson, A. S., & Henkel, C. (1997). A Survey for H_2O Megamasers in Active Galactic Nuclei. II. A Comparison of Detected and Undetected Galaxies. *ApJS*, 110, 321–+. 13, 22, 86, 96, 97

Braine, J., Combes, F., Casoli, F., Dupraz, C., Gerin, M., Klein, U., Wielebinski, R., & Brouillet, N. (1993). A CO(1-0) and CO(2-1) survey of nearby spiral galaxies. I - Data and observations. *A&AS*, 97, 887–936. 70

Bronfman, L., Nyman, L.-A., & May, J. (1996). A CS(2-1) survey of IRAS point sources with color characteristics of ultra-compact HII regions. *A&AS*, 115, 81–+. 88

Cao, C., Xia, X. Y., Wu, H., Mao, S., Hao, C. N., & Deng, Z. G. (2008). Mid-Infrared spectroscopic properties of ultra-luminous infrared quasars. *MNRAS*, 390, 336–348. 15

Casoli, F. & Gerin, M. (1993). CO in the 'Black Eye' galaxy NGC 4826. *A&A*, 279, L41–L44. 70

Castangia, P., Tarchi, A., Henkel, C., & Menten, K. M. (2008). New H_2O masers in Seyfert and FIR bright galaxies II. The intermediate luminosity range. *A&A*, 479, 111–122. 91, 94

Churchwell, E., Witzel, A., Huchtmeier, W., Pauliny-Toth, I., Roland, J., & Sieber, W. (1977). Detection of H_2O maser emission in the Galaxy M 33. *A&A*, 54, 969–971. 21

Colina, L., Arribas, S., & Clements, D. (2004). INTEGRAL Field Spectroscopy of the Extended Ionized Gas in Arp 220. *ApJ*, 602, 181–189. 41, 44

Combes, F. (2006). Starbursts and AGN Fueling through Secular Evolution. In *Revista Mexicana de Astronomia y Astrofisica Conference Series*, volume 26 of *Revista Mexicana de Astronomia y Astrofisica, vol. 27* (pp. 131–134). 98

Combes, F., García-Burillo, S., Boone, F., Hunt, L. K., Baker, A. J., Eckart, A., Englmaier, P., Leon, S., Neri, R., Schinnerer, E., & Tacconi, L. J.

(2004). Molecular gas in NUclei of GAlaxies (NUGA). II. The ringed LINER NGC 7217. *A&A*, 414, 857–872. 70

Condon, J. J., Cotton, W. D., Greisen, E. W., Yin, Q. F., Perley, R. A., Taylor, G. B., & Broderick, J. J. (1998). The NRAO VLA Sky Survey. *AJ*, 115, 1693–1716. 82

de Vaucouleurs, G. (1953). On the distribution of mass and luminosity in elliptical galaxies. *MNRAS*, 113, 134–+. 43

Desert, F. X. (1986). Infrared Spectra and Dust Temperature Fluctuations. In F. P. Israel (Ed.), *LIGHT ON DARK MATTER: 1ST IRAS CONF. 1986 P. 213, 1986* (pp. 213–+). 78, 79

Downes, D. & Eckart, A. (2007). Black hole in the West nucleus of Arp 220. *A&A*, 468, L57–L61. 36, 38, 39, 41, 42, 43, 52

Downes, D. & Solomon, P. M. (1998). Rotating Nuclear Rings and Extreme Starbursts in Ultraluminous Galaxies. *ApJ*, 507, 615–654. 37, 39, 44

Dressler, A. (1980). Galaxy morphology in rich clusters - Implications for the formation and evolution of galaxies. *ApJ*, 236, 351–365. 74

Dressler, A., Oemler, A. J., Couch, W. J., Smail, I., Ellis, R. S., Barger, A., Butcher, H., Poggianti, B. M., & Sharples, R. M. (1997). Evolution since $Z = 0.5$ of the Morphology-Density Relation for Clusters of Galaxies. *ApJ*, 490, 577–+. 74

Dressler, A., Thompson, I. B., & Shectman, S. A. (1985). Statistics of emission-line galaxies in rich clusters. *ApJ*, 288, 481–486. 74

Eckart, A. & Downes, D. (2001). Arp 220: A Circumnuclear Polar Ring as an Alternative to a Double Nucleus? *ApJ*, 551, 730–742. 43, 46, 53

Evans, A. S., Frayer, D. T., Surace, J. A., & Sanders, D. B. (2001). Erratum: "Molecular Gas in Infrared-Excess, Optically Selected Quasars and the Connection with Infrared-luminous Galaxies" [Astron. J. 121, 1893 (2001)]. *AJ*, 121, 3285–3285. 20

Evans, A. S., Solomon, P. M., Tacconi, L. J., Vavilkin, T., & Downes, D. (2006). Dense Molecular Gas and the Role of Star Formation in the Host Galaxies of Quasi-stellar Objects. *AJ*, 132, 2398–2408. 20

Ewen, H. I. & Purcell, E. M. (1951). Observation of a Line in the Galactic Radio Spectrum: Radiation from Galactic Hydrogen at 1,420 Mc./sec. *Nature*, 168, 356–+. 16

Farrah, D., Rowan-Robinson, M., Oliver, S., Serjeant, S., Borne, K., Lawrence, A., Lucas, R. A., Bushouse, H., & Colina, L. (2001). HST/WFPC2 imaging of the QDOT ultraluminous infrared galaxy sample. *MNRAS*, 326, 1333–1352. 14

Forbrich, J. (2003). Infrarot– und Submilimeteruntersuchungen zur Entstehung massereicher Sterne. Diplomarbeit, Friedrich–Schiller–Universität Jena, Physikalisch–Astronomische Fakultät. 19

Förster Schreiber, N. M., Genzel, R., Lehnert, M. D., Bouché, N., Verma, A., Erb, D. K., Shapley, A. E., Steidel, C. C., Davies, R., Lutz, D., Nesvadba, N., Tacconi, L. J., Eisenhauer, F., Abuter, R., Gilbert, A., Gillessen, S., & Sternberg, A. (2006). SINFONI Integral Field Spectroscopy of z ~ 2 UV-selected Galaxies: Rotation Curves and Dynamical Evolution. *ApJ*, 645, 1062–1075. 36

Gao, Y. & Solomon, P. M. (2004). The Star Formation Rate and Dense Molecular Gas in Galaxies. *ApJ*, 606, 271–290. 14

Gebhardt, K., Bender, R., Bower, G., Dressler, A., Faber, S. M., Filippenko, A. V., Green, R., Grillmair, C., Ho, L. C., Kormendy, J., Lauer, T. R., Magorrian, J., Pinkney, J., Richstone, D., & Tremaine, S. (2000). A Relationship between Nuclear Black Hole Mass and Galaxy Velocity Dispersion. *ApJ*, 539, L13–L16. 8

Ghosh, K. K., Ramsey, B. D., Sadun, A. C., & Soundararajaperumal, S. (2000). Optical Variability of Blazars. *ApJS*, 127, 11–26. 10

Goldschmidt, P., Kukula, M. J., Miller, L., & Dunlop, J. S. (1999). A Comparison of the Optical Properties of Radio-loud and Radio-quiet Quasars. *ApJ*, 511, 612–624. 10

Goldsmith, P. F., Pandian, J. D., & Deshpande, A. A. (2008). A Search for 6.7 GHz Methanol Masers in M33. *ApJ*, 680, 1132–1136. 89

González-García, A. C. & Balcells, M. (2005). Elliptical galaxies from mergers of discs. *MNRAS*, 357, 753–772. 48

Greene, J. E. & Ho, L. C. (2006a). Measuring Stellar Velocity Dispersions in Active Galaxies. *ApJ*, 641, 117–132. 88

Greene, J. E. & Ho, L. C. (2006b). The M_{BH}-σ_* Relation in Local Active Galaxies. *ApJ*, 641, L21–L24. 88

Greenhill, L. J. (2007). Masers in AGN environments. In J. M. Chapman & W. A. Baan (Eds.), *IAU Symposium*, volume 242 of *IAU Symposium* (pp. 381–390). 89

Greenhill, L. J., Gwinn, C. R., Antonucci, R., & Barvainis, R. (1996). VLBI Imaging of Water Maser Emission from the Nuclear Torus of NGC 1068. *ApJ*, 472, L21+. 97

Greenhill, L. J., Kondratko, P. T., Lovell, J. E. J., Kuiper, T. B. H., Moran, J. M., Jauncey, D. L., & Baines, G. P. (2003). The Discovery of H_2O Maser Emission in Seven Active Galactic Nuclei and at High Velocities in the Circinus Galaxy. *ApJ*, 582, L11–L14. 93

Greenstein, J. L. & Schmidt, M. (1964). The Quasi-Stellar Radio Sources 3c 48 and 3c 273. *ApJ*, 140, 1–+. 7

Haan, S., Schinnerer, E., Mundell, C. G., García-Burillo, S., & Combes, F. (2008). Atomic Hydrogen Properties of Active Galactic Nuclei Host Galaxies: H I in 16 Nuclei of Galaxies (nuga) Sources. *AJ*, 135, 232–257. 17, 18, 65, 69, 70, 73, 76, 77, 83, 137

Hagiwara, Y., Henkel, C., Menten, K. M., & Nakai, N. (2001). Water Maser Emission from the Active Nucleus in M51. *ApJ*, 560, L37–L40. 22

Hartmann, D. & Burton, W. B. (1997). *Atlas of Galactic Neutral Hydrogen.* Atlas of Galactic Neutral Hydrogen, by Dap Hartmann and W. Butler Burton, pp. 243. ISBN 0521471117. Cambridge, UK: Cambridge University Press, February 1997. 68, 83

Heckman, T. M. (1980). An optical and radio survey of the nuclei of bright galaxies - Activity in normal galactic nuclei. *A&A*, 87, 152–164. 13

Helfer, T. T., Thornley, M. D., Regan, M. W., Wong, T., Sheth, K., Vogel, S. N., Blitz, L., & Bock, D. C.-J. (2003). The BIMA Survey of Nearby Galaxies (BIMA SONG). II. The CO Data. *ApJS*, 145, 259–327. 70

Helou, G. (1986). The IRAS colors of normal galaxies. *ApJ*, 311, L33–L36. 78

Henkel, C., Braatz, J. A., Greenhill, L. J., & Wilson, A. S. (2002). Discovery of water vapor megamaser emission from Mrk 1419 (NGC 2960): An analogue of NGC 4258? *A&A*, 394, L23–L26. 98

Henkel, C., Peck, A. B., Tarchi, A., Nagar, N. M., Braatz, J. A., Castangia, P., & Moscadelli, L. (2005). New H_2O masers in Seyfert and FIR bright galaxies. *A&A*, 436, 75–90. 20, 85

Herrnstein, J. R., Moran, J. M., Greenhill, L. J., Diamond, P. J., Inoue, M., Nakai, N., Miyoshi, M., Henkel, C., & Riess, A. (1999). A geometric distance to the galaxy NGC4258 from orbital motions in a nuclear gas disk. *Nature*, 400, 539–541. 85, 98

Hewitt, J. N., Haynes, M. P., & Giovanelli, R. (1983). Neutral hydrogen in isolated galaxies. II - The large angular diameter galaxies. *AJ*, 88, 272–295. 17

Hibbard, J. E., Vacca, W. D., & Yun, M. S. (2000). The Neutral Hydrogen Distribution in Merging Galaxies: Differences between Stellar and Gaseous Tidal Morphologies. *AJ*, 119, 1130–1144. 55

Ho, L. C., Darling, J., & Greene, J. E. (2008a). A New H I Survey of Active Galaxies. *ApJS*, 177, 103–130. 17, 58, 60, 68, 69

Ho, L. C., Darling, J., & Greene, J. E. (2008b). Properties of Active Galaxies Deduced from H I Observations. *ApJ*, 681, 128–140. 17, 58, 69

Ho, L. C., Filippenko, A. V., & Sargent, W. L. W. (1993). A Reevaluation of the Excitation Mechanism of LINERs. *ApJ*, 417, 63–+. 13

Hopkins, P. F., Hernquist, L., Martini, P., Cox, T. J., Robertson, B., Di Matteo, T., & Springel, V. (2005). A Physical Model for the Origin of Quasar Lifetimes. *ApJ*, 625, L71–L74. 20

Hou, L. G., Wu, X.-B., & Han, J. L. (2009). Ultra-luminous Infrared galaxies in SDSS Data Release 6. *ArXiv e-prints*. 14

Impellizzeri, C. M. V., McKean, J. P., Castangia, P., Roy, A. L., Henkel, C., Brunthaler, A., & Wucknitz, O. (2008). A gravitationally lensed water maser in the early Universe. *Nature*, 456, 927–929. 86, 96

Joseph, R. D. (1999). The great debate: Starbusts as the energy source of ultraluminous infrared galaxies. *Ap&SS*, 266, 321–329. 14, 35

Kawakatu, N., Anabuki, N., Nagao, T., Umemura, M., Nakagawa, T., & Mori, M. (2006). Formation of SMBHs and QSO evolution. *New Astronomy Review*, 50, 769–771. 15

Kohno, K., Kawabe, R., & Vila-Vilaró, B. (1999). Dense Molecular Gas Associated with the Circumnuclear Star-forming Ring in the Barred Spiral Galaxy NGC 6951. *ApJ*, 511, 157–177. 70

Kondratko, P. T., Greenhill, L. J., & Moran, J. M. (2006a). Discovery of Water Maser Emission in Five AGNs and a Possible Correlation Between Water Maser and Nuclear 2-10 keV Luminosities. *ApJ*, 652, 136–145. 89, 93

Kondratko, P. T., Greenhill, L. J., Moran, J. M., Lovell, J. E. J., Kuiper, T. B. H., Jauncey, D. L., Cameron, L. B., Gómez, J. F., García-Miró, C., Moll, E., de Gregorio-Monsalvo, I., & Jiménez-Bailón, E. (2006b). Discovery of Water Maser Emission in Eight AGNs with 70 m Antennas of NASA's Deep Space Network. *ApJ*, 638, 100–105. 13, 22, 86, 93, 97

Kormendy, J. & Gebhardt, K. (2001). Supermassive black holes in galactic nuclei. In J. C. Wheeler & H. Martel (Ed.), *20th Texas Symposium on relativistic astrophysics*, volume 586 of *American Institute of Physics Conference Series* (pp. 363–381). 8

Krips, M., Eckart, A., Neri, R., Bertram, T., Straubmeier, C., Fischer, S., Staguhn, J. G., & Vogel, S. N. (2007). Barred CO emission in HE 1029-1831. *A&A*, 464, 187–191. 20

Krips, M., Eckart, A., Neri, R., Schödel, R., Leon, S., Downes, D., García-Burillo, S., & Combes, F. (2006a). Continuum emission in NGC 1068 and NGC 3147: indications for a turnover in the core spectra. *A&A*, 446, 113–120. 19, 58

Krips, M., Neri, R., Garcia-Burillo, S., Combes, F., Martin, S., Eckart, A., Petitpas, G., & Peck, A. (2006b). A HCN and HCO+ Multi-transition Line Survey in Active Galaxies: AGN versus Starburst Environments. In *Bulletin of the American Astronomical Society*, volume 38 of *Bulletin of the American Astronomical Society* (pp. 1060–+). 19, 58

Krolik, J. H. & Kallman, T. R. (1987). Fe K features as probes of the nuclear reflection region in Seyfert galaxies. *ApJ*, 320, L5–L8. 11

Kuo, C., Lim, J., Tang, Y., & Ho, P. T. P. (2008). Prevalence of Tidal Interactions among Local Seyfert Galaxies. *ApJ*, 679, 1047–1093. 17

Labita, M., Treves, A., Falomo, R., & Uslenghi, M. (2006). The BH mass of nearby QSOs: a comparison of the bulge luminosity and virial methods. *MNRAS*, 373, 551–560. 98

Lawrence, A. & Elvis, M. (1982). Obscuration and the various kinds of Seyfert galaxies. *ApJ*, 256, 410–426. 96

Lim, J. & Ho, P. T. P. (1999). Violent Tidal Disruptions of Atomic Hydrogen Gas in Quasar Host Galaxies. *ApJ*, 510, L7–L10. 17, 58

Lim, J., Ho, P. T. P., Hua-Ting, C., Shyang, W., & Wen-Shuo, L. (2001). H I Imaging of Low-Redshift QSO Host Galaxies. In J. E. Hibbard, M. Rupen,

& J. H. van Gorkom (Eds.), *Gas and Galaxy Evolution*, volume 240 of *Astronomical Society of the Pacific Conference Series* (pp. 111–+). 17, 58

Lo, K. Y. (2005). Mega-Masers and Galaxies. *ARA&A*, 43, 625–676. 21, 85

Lonsdale, C. J., Farrah, D., & Smith, H. E. (2006). *Ultraluminous Infrared Galaxies*, (pp. 285–+). Springer Verlag. 68

Madejski, G., Done, C., Życki, P. T., & Greenhill, L. (2006). X-Ray Emission from Megamaser Galaxy IC 2560. *ApJ*, 636, 75–82. 97

Magorrian, J., Tremaine, S., Richstone, D., Bender, R., Bower, G., Dressler, A., Faber, S. M., Gebhardt, K., Green, R., Grillmair, C., Kormendy, J., & Lauer, T. (1998). The Demography of Massive Dark Objects in Galaxy Centers. *AJ*, 115, 2285–2305. 7

Maiolino, R., Ruiz, M., Rieke, G. H., & Papadopoulos, P. (1997). Molecular Gas, Morphology, and Seyfert Galaxy Activity. *ApJ*, 485, 552–+. 69

Martin, M. C. (1998). Catalogue of HI maps of galaxies. II. Analysis of the data. *A&AS*, 131, 77–87. 17

Martini, P., Kelson, D. D., Kim, E., Mulchaey, J. S., & Athey, A. A. (2006). Spectroscopic Confirmation of a Large Population of Active Galactic Nuclei in Clusters of Galaxies. *ApJ*, 644, 116–132. 74

Matveenko, L. I., Diamond, P. J., & Graham, D. A. (2000). Ring Structures in Orion KL. *Astronomy Reports*, 44, 592–610. 88

Mauersberger, R., Henkel, C., & Wilson, T. L. (1987). A multilevel study of ammonia in star-forming regions. I - Maser and thermal emission toward W51 IRS 2. *A&A*, 173, 352–360. 87

Mirabel, I. F. & Sanders, D. B. (1988). 21 centimeter survey of luminous infrared galaxies. *ApJ*, 335, 104–121. 68

Miyoshi, M., Moran, J., Herrnstein, J., Greenhill, L., Nakai, N., Diamond, P., & Inoue, M. (1995). Evidence for a Black-Hole from High Rotation Velocities in a Sub-Parsec Region of NGC4258. *Nature*, 373, 127–+. 85, 96, 97

Moshir, M., Kopan, G., Conrow, T., McCallon, H., Hacking, P., Gregorich, D., Rohrbach, G., Melnyk, M., Rice, W., Fullmer, L., White, J., & Chester, T. (1990). The IRAS Faint Source Catalog, Version 2. In *Bulletin of the American Astronomical Society*, volume 22 of *Bulletin of the American Astronomical Society* (pp. 1325–+). 65

Mundell, C. G., Ferruit, P., & Pedlar, A. (2001). Nuclear Gasdynamics in Arp 220: Subkiloparsec-Scale Atomic Hydrogen Disks. *ApJ*, 560, 168–177. 49, 52, 56

Nardini, E., Risaliti, G., Salvati, M., Sani, E., Imanishi, M., Marconi, A., & Maiolino, R. (2008). Spectral decomposition of starbursts and active galactic nuclei in 5-8 μm Spitzer-IRS spectra of local ultraluminous infrared galaxies. *MNRAS*, 385, L130–L134. 14

Nishiyama, K. & Nakai, N. (2001). CO Survey of Nearby Spiral Galaxies with the Nobeyama 45-m Telescope: I. The Data. *PASJ*, 53, 713–756. 70

Norris, R. P. (1988). The double radio nucleus of ARP 220. *MNRAS*, 230, 345–351. 35

Omont, A. (2007). Molecules in galaxies. *Reports on Progress in Physics*, 70, 1099–1176. 18, 19

O'Neill, P. M., Nandra, K., Papadakis, I. E., & Turner, T. J. (2005). The relationship between X-ray variability amplitude and black hole mass in active galactic nuclei. *MNRAS*, 358, 1405–1416. 88, 99

Ott, M., Witzel, A., Quirrenbach, A., Krichbaum, T. P., Standke, K. J., Schalinski, C. J., & Hummel, C. A. (1994). An updated list of radio flux density calibrators. *A&A*, 284, 331–339. 60, 87

Parra, R., Conway, J. E., Diamond, P. J., Thrall, H., Lonsdale, C. J., Lonsdale, C. J., & Smith, H. E. (2007). The Radio Spectra of the Compact Sources in Arp 220: A Mixed Population of Supernovae and Supernova Remnants. *ApJ*, 659, 314–330. 36

Peterson, B. M. (1997). *An Introduction to Active Galactic Nuclei*. 10

Pimbblet, K. A. (2003). At the Vigintennial of the Butcher-Oemler Effect. *Publications of the Astronomical Society of Australia*, 20, 294–299. 74

Popesso, P. & Biviano, A. (2006). The AGN fraction-velocity dispersion relation in clusters of galaxies. *A&A*, 460, L23–L26. 74

Pott, J.-U., Hartwich, M., Eckart, A., Leon, S., Krips, M., & Straubmeier, C. (2004). Warped molecular gas disk in NGC 3718. *A&A*, 415, 27–38. 70

Richstone, D., Gebhardt, K., & Pinkney, J. (1999). Black Holes in Nearby Galaxies. In *Bulletin of the American Astronomical Society*, volume 31 of *Bulletin of the American Astronomical Society* (pp. 1518–+). 8

Rieke, G. H. & Lebofsky, M. J. (1985). The interstellar extinction law from 1 to 13 microns. *ApJ*, 288, 618–621. 40

Rieke, G. H. & Low, F. J. (1972). Infrared Photometry of Extragalactic Sources. *ApJ*, 176, L95+. 13

Romano, D., Matteucci, F., Salucci, P., & Chiappini, C. (2000). The Mass Surface Density in the Local Disk and the Chemical Evolution of the Galaxy. *ApJ*, 539, 235–240. 77

Sage, L. J. (1993). Molecular Gas in Nearby Galaxies - Part One - Co/ Observations of a Distance-Limited Sample. *A&A*, 272, 123–+. 70

Sakai, S., Mould, J. R., Hughes, S. M. G., Huchra, J. P., Macri, L. M., Kennicutt, Jr., R. C., Gibson, B. K., Ferrarese, L., Freedman, W. L., Han, M., Ford, H. C., Graham, J. A., Illingworth, G. D., Kelson, D. D., Madore, B. F., Sebo, K., Silbermann, N. A., & Stetson, P. B. (2000). The Hubble Space Telescope Key Project on the Extragalactic Distance Scale. XXIV. The Calibration of Tully-Fisher Relations and the Value of the Hubble Constant. *ApJ*, 529, 698–722. 77, 78

Sakamoto, K., Okumura, S., Minezaki, T., Kobayashi, Y., & Wada, K. (1995). Bar-Driven Gas Structure and Star Formation in the Center of M100. *AJ*, 110, 2075–+. 70

Sakamoto, K., Scoville, N. Z., Yun, M. S., Crosas, M., Genzel, R., & Tacconi, L. J. (1999). Counterrotating Nuclear Disks in ARP 220. *ApJ*, 514, 68–76. 36, 39, 43, 52

Sakamoto, K., Wang, J., Wiedner, M. C., Wang, Z., Peck, A. B., Zhang, Q., Petitpas, G. R., Ho, P. T. P., & Wilner, D. J. (2008). Submillimeter Array Imaging of the CO(3-2) Line and 860 μm Continuum of Arp 220: Tracing the Spatial Distribution of Luminosity. *ApJ*, 684, 957–977. 35

Sanders, D. B. (1999). The "Great Debate": The case for AGNs. *Ap&SS*, 266, 331–348. 14, 35

Sanders, D. B. & Mirabel, I. F. (1996). Luminous Infrared Galaxies. *ARA&A*, 34, 749–+. 65

Sanders, D. B., Scoville, N. Z., & Soifer, B. T. (1991). Molecular gas in luminous infrared galaxies. *ApJ*, 370, 158–171. 14

Sanders, D. B., Soifer, B. T., Elias, J. H., Madore, B. F., Matthews, K., Neugebauer, G., & Scoville, N. Z. (1988). Ultraluminous infrared galaxies and the origin of quasars. *ApJ*, 325, 74–91. 1, 10, 14, 15, 35, 57, 58

Schmidt, M. (1963). 3C 273 : A Star-Like Object with Large Red-Shift. *Nature*, 197, 1040–+. 7

Schmidt, M. (1988). Space Distribution and Luminosity Function of Quasars. In H. R. Miller & P. J. Wiita (Ed.), *Active Galactic Nuclei*, volume 307 of *Lecture Notes in Physics, Berlin Springer Verlag* (pp. 408–+). 10

Schöniger, F. & Sofue, Y. (1997). The CO Tully-Fisher relation for the Virgo cluster. *A&A*, 323, 14–20. 77

Scoville, N. Z. (2000). Ultra-Luminous IR Galaxies at Low and High Redshift. In F. Combes, G. A. Mamon, & V. Charmandaris (Eds.), *Dynamics of Galaxies: from the Early Universe to the Present*, volume 197 of *Astronomical Society of the Pacific Conference Series* (pp. 301–+). 36

Scoville, N. Z., Evans, A. S., Dinshaw, N., Thompson, R., Rieke, M., Schneider, G., Low, F. J., Hines, D., Stobie, B., Becklin, E., & Epps, H. (1998). NICMOS Imaging of the Nuclei of ARP 220. *ApJ*, 492, L107+. 37, 38

Scoville, N. Z., Frayer, D. T., Schinnerer, E., & Christopher, M. (2003). The Host Galaxies of Optically Bright Quasi-stellar Objects: Molecular Gas in $z \leq 0.1$ Palomar-Green Quasi-stellar Objects. *ApJ*, 585, L105–L108. 19, 20, 57

Scoville, N. Z., Yun, M. S., & Bryant, P. M. (1997). Arcsecond Imaging of CO Emission in the Nucleus of ARP 220. *ApJ*, 484, 702–+. 36, 40, 42, 43, 46, 52

Shields, G. A., Gebhardt, K., Salviander, S., Wills, B. J., Xie, B., Brotherton, M. S., Yuan, J., & Dietrich, M. (2003). The Black Hole-Bulge Relationship in Quasars. *ApJ*, 583, 124–133. 8

Shostak, G. S. (1978). Integral properties of late-type galaxies derived from H I observations. *A&A*, 68, 321–341. 64

Sofue, Y., Koda, J., Nakanishi, H., Onodera, S., Kohno, K., Tomita, A., & Okumura, S. K. (2003). The Virgo High-Resolution CO Survey: I. CO Atlas. *PASJ*, 55, 17–58. 70

Soifer, B. T., Sanders, D. B., Madore, B. F., Neugebauer, G., Danielson, G. E., Elias, J. H., Lonsdale, C. J., & Rice, W. L. (1987). The IRAS bright galaxy sample. II - The sample and luminosity function. *ApJ*, 320, 238–257. 35

Soifer, B. T., Sanders, D. B., Neugebauer, G., Danielson, G. E., Lonsdale, C. J., Madore, B. F., & Persson, S. E. (1986). The luminosity function and space density of the most luminous galaxies in the IRAS survey. *ApJ*, 303, L41–L44. 14

Springob, C. M., Haynes, M. P., Giovanelli, R., & Kent, B. R. (2005). A Digital Archive of H I 21 Centimeter Line Spectra of Optically Targeted Galaxies. *ApJS*, 160, 149–162. 60

Staguhn, J. G., Schinnerer, E., Eckart, A., & Scharwächter, J. (2004). Resolving the Host Galaxy of the Nearby QSO I Zw 1 with Subarcsecond Multitransition Molecular Line Observations. *ApJ*, 609, 85–93. 20

Taniguchi, Y., Ikeuchi, S., & Shioya, Y. (1999). Formation of Quasar Nuclei in the Hearts of Ultraluminous Infrared Galaxies. *ApJ*, 514, L9–L12. 35

Tarchi, A., Brunthaler, A., Henkel, C., Menten, K. M., Braatz, J., & Weiß, A. (2007). The innermost region of the water megamaser radio galaxy 3C 403. *A&A*, 475, 497–506. 99

Taylor, G. B., Carilli, C. L., & Perley, R. A., Eds. (1999). *Synthesis Imaging in Radio Astronomy II*, volume 180 of *Astronomical Society of the Pacific Conference Series*. 28

Terlevich, R., Tenorio-Tagle, G., Franco, J., & Melnick, J. (1992). The starburst model for active galactic nuclei - The broad-line region as supernova remnants evolving in a high-density medium. *MNRAS*, 255, 713–728. 8

Tifft, W. G. & Huchtmeier, W. K. (1990). Comparisons between 21 CM data from Effelsberg and Greenbank. *A&AS*, 84, 47–58. 60

Toomre, A. & Toomre, J. (1972). Galactic Bridges and Tails. *ApJ*, 178, 623–666. 46

Tutui, Y. & Sofue, Y. (1997). Effects of galaxy interaction on the Tully-Fisher relation: CO VS HI linewidths. *A&A*, 326, 915–918. 70

Urry, C. M. & Padovani, P. (1995). Unified Schemes for Radio-Loud Active Galactic Nuclei. *PASP*, 107, 803–+. 9, 12

Veilleux, S., Kim, D.-C., & Sanders, D. B. (2002). Optical and Near-Infrared Imaging of the IRAS 1 Jy Sample of Ultraluminous Infrared Galaxies. II. The Analysis. *ApJS*, 143, 315–376. 14

Veilleux, S., Rupke, D. S. N., Kim, D.-C., Genzel, R., Sturm, E., Lutz, D., Contursi, A., Schweitzer, M., Tacconi, L. J., Netzer, H., Sternberg, A., Mihos, J. C., Baker, A. J., Mazzarella, J. M., Lord, S., Sanders, D. B., Stockton, A., Joseph, R. D., & Barnes, J. E. (2009). Spitzer Quasar and

Ulirg Evolution Study (QUEST). IV. Comparison of 1 Jy Ultraluminous Infrared Galaxies with Palomar-Green Quasars. *ApJS*, 182, 628–666. 15

Vestergaard, M., Fan, X., Tremonti, C. A., Osmer, P. S., & Richards, G. T. (2008). Mass Functions of the Active Black Holes in Distant Quasars from the Sloan Digital Sky Survey Data Release 3. *ApJ*, 674, L1–L4. 99

Wang, T. & Lu, Y. (2001). Black hole mass and velocity dispersion of narrow line region in active galactic nuclei and narrow line Seyfert 1 galaxies. *A&A*, 377, 52–59. 88, 99

Wilkes, B. J., Schmidt, G. D., Smith, P. S., Mathur, S., & McLeod, K. K. (1995). Optical Detection of the Hidden Nuclear Engine in NGC 4258. *ApJ*, 455, L13+. 12

Wisotzki, L., Christlieb, N., Bade, N., Beckmann, V., Köhler, T., Vanelle, C., & Reimers, D. (2000). The Hamburg/ESO survey for bright QSOs. III. A large flux-limited sample of QSOs. *A&A*, 358, 77–87. 58

Wright, G. S., James, P. A., Joseph, R. D., & McLean, I. S. (1990). Elliptical-like light profiles in infrared images of merging spiral galaxies. *Nature*, 344, 417–419. 43

Young, J. S., Xie, S., Kenney, J. D. P., & Rice, W. L. (1989). Global properties of infrared bright galaxies. *ApJS*, 70, 699–722. 69

Young, J. S., Xie, S., Tacconi, L., Knezek, P., Viscuso, P., Tacconi-Garman, L., Scoville, N., Schneider, S., Schloerb, F. P., Lord, S., Lesser, A., Kenney, J., Huang, Y.-L., Devereux, N., Claussen, M., Case, J., Carpenter, J., Berry, M., & Allen, L. (1995). The FCRAO Extragalactic CO Survey. I. The Data. *ApJS*, 98, 219–+. 70

Zhang, J. S., Henkel, C., Kadler, M., Greenhill, L. J., Nagar, N., Wilson, A. S., & Braatz, J. A. (2006). Extragalactic H_2O masers and X-ray absorbing column densities. *A&A*, 450, 933–944. 97

Ziegler, B. L., Böhm, A., Fricke, K. J., Jäger, K., Nicklas, H., Bender, R., Drory, N., Gabasch, A., Saglia, R. P., Seitz, S., Heidt, J., Mehlert, D.,

Möllenhoff, C., Noll, S., & Sutorius, E. (2002). The Evolution of the Tully-Fisher Relation of Spiral Galaxies. *ApJ*, 564, L69–L72. 77, 78

List of Figures

1.1.	The type 1 QSO NGC 4593	9
1.2.	The blazar 3C 279	10
1.3.	The type 1 Seyfert galaxy NGC 7742	11
1.4.	The unified scheme	12
1.5.	The LINER galaxy M 51	13
1.6.	The ULIRG Arp 299	14
1.7.	NGC 4321 in HI and ^{12}CO$(3-2)$	18
2.1.	A schematic diagram of a single-dish telescope	25
2.2.	A single-dish telescope and its main characteristics	26
2.3.	The Effelsberg 100-m telescope	27
2.4.	A schematic interferometer diagram	28
2.5.	The Very Large Array	30
3.1.	Arp 220 at different scales and wavelengths	38
3.2.	CO$(2-1)$ p-v diagram, velocity field and velocity distribution of Arp 220	40
3.3.	CO$(1-0)$ velocity field of Arp 220	42
3.4.	CO$(1-0)$ intensity map of Arp 220	44
3.5.	The Identikit model parameter space	46
3.6.	The best model fit for Arp 220	47
3.7.	Arp 220 HST image overlay with the best model	49
3.8.	Comparison of the p-v diagrams	50
3.9.	Model evaluation criteria	51
3.10.	Counterexamples to the best model fit	53
4.1.	Spectra and DSS images of the detected sources	61

List of Figures

4.2.	HI mass vs. infrared luminosity	68
4.3.	HI flux vs. CO intensity	71
4.4.	Infrared luminosity vs. compactness	75
4.5.	Tully-Fisher diagram	77
4.6.	IRAS color-color diagram	78
4.7.	HI continuum and line maps of HE 1248−1356	82
5.1.	Orion–KL and HE 0119−0118 in H_2O	90
5.2.	The averaged water maser spectrum	93
5.3.	The sensitivity of the observations at hand	94
5.4.	Atomic and molecular gas properties of the QSOs	95
5.5.	The unified scheme in the context of H_2O megamasers ...	97
A.1.	H_2O spectra and DSS images of the sources searched for maser emission	102
A.2.	H_2O spectra and cross-correlation functions of HE 0021−1819 and HE 0040−1105	108
A.3.	H_2O spectra and cross-correlation functions of HE 0114−0015 and HE 0150−0344	109
A.4.	H_2O spectra and cross-correlation functions of HE 0212−0059 and HE 0224−2834	110
A.5.	H_2O spectra and cross-correlation functions of HE 0232−0900 and HE 0345+0056	111
A.6.	H_2O spectra and cross-correlation functions of HE 0433−1028 and HE 0853−0126	112
A.7.	H_2O spectra and cross-correlation functions of HE 1011−0403 and HE 1017−0305	113
A.8.	H_2O spectra and cross-correlation functions of HE 1107−0813 and HE 1126−0407	114
A.9.	H_2O spectra and cross-correlation functions of HE 2233+0124 and HE 2302−0857	115
A.10.	H_2O spectra and cross-correlation functions of W3(OH) ..	116

List of Tables

4.1.	List of sources observed at 21 cm	59
4.2.	Summary of the HI properties	66
4.3.	Sources from Haan et al. (2008)	70
4.4.	Kendall-τ test parameter overview	80
5.1.	Sources searched for H_2O maser emission	88

Acknowledgements

First of all, I'd like to thank Prof. Dr. Andreas Eckart for giving me the opportunity to write this thesis under his supervision. I learned a lot from the experience.

Additionally, I want to thank Prof. Dr. Andreas Zilges for taking over the task as second referee, and Prof. Dr. Joachim Saur and Dr. Steffen Rost for completing my thesis commitee.

Furthermore, I owe thanks to Dr. Dennis Downes for providing the PdBI data for Arp 220.

Very special thanks go to Macarena García-Marín and Thomas Bertram, who where always willing to lend me an ear for my extragalactic questions. I'm very grateful to Melanie Krips for always answering any stupid question I could come up with, regarding interferometry and data reduction. My thanks also go to Thomas Bertram, Jens Zuther and Andreas Breslau who helped me with any computer related problem that came up. I'm very grateful for the nice and relaxed working atmosphere in the 'aegroup'. Especially Devaky Kunneriath, Nadeen Sabha and Monica Valencia-Schneider who welcomed me to their office and endured my presence during the last year of my thesis work. I very much appreciated all the 'non-working' related discussions with people from the 'OoI' , especially Imke Wank and Lydia Moser. They very much helped to get my head clear again.

Additionally I thank all the people making the effort to read my thesis: Macarena García-Marín, Christina Hövel, Thomas Kaczmarek and Jens Zuther.

Acknowledgements

And last, but not least at all, the biggest thank you goes to my family. Though it was not easy in the last months; loosing never is gaining. Their strength, patience and encouragement made me who I am today.

Die VDM Verlagsservicegesellschaft sucht für wissenschaftliche Verlage abgeschlossene und herausragende

Dissertationen, Habilitationen, Diplomarbeiten, Master Theses, Magisterarbeiten usw.

für die kostenlose Publikation als Fachbuch.

Sie verfügen über eine Arbeit, die hohen inhaltlichen und formalen Ansprüchen genügt, und haben Interesse an einer honorarvergüteten Publikation?

Dann senden Sie bitte erste Informationen über sich und Ihre Arbeit per Email an *info@vdm-vsg.de*.

Sie erhalten kurzfristig unser Feedback!

VDM Verlagsservicegesellschaft mbH
Dudweiler Landstr. 99   Telefon +49 681 3720 174
D - 66123 Saarbrücken   Fax   +49 681 3720 1749
www.vdm-vsg.de

Die VDM Verlagsservicegesellschaft mbH vertritt

Printed by Books on Demand GmbH, Norderstedt / Germany